Handbook of polymer composites for engineers

Edited by LEONARD HOLLAWAY

Handbook of polymer composites for engineers

**Published in association with the
British Plastics Federation**

WOODHEAD PUBLISHING LIMITED
Cambridge England

Published by Woodhead Publishing Ltd,
Abington Hall, Abington,
Cambridge CB1 6AH, England

First published 1994, Woodhead Publishing Ltd

British Library Cataloguing in Publication Data
A catalogue record for this book is available from the British Library

ISBN 1 85573 129 0

Designed by Geoff Green (text) and Chris Feely (jacket)
Typeset by BookEns Ltd., Baldock, Herts.
Printed by St Edmundsbury Press, Suffolk, England

Contents

Preface   *ix*
List of contributors   *xi*

Part I Introduction   1

1   Fibre reinforced polymers   3
     LESLIE S. NORWOOD
     1.1   History of fibre reinforced polymer materials   3
     1.2   Resins   5
     1.3   Additives to resins   21
     1.4   Reinforcements   23
     1.5   Properties of composite materials   33
     1.6   Health and safety   63

Part II Techniques for manufacture of composites   71

2   Manufacturing processes   73
     PHILIP BALL
     2.1   Introduction   73
     2.2   Open mould processes   74
     2.3   Closed mould processes   83
     2.4   Continuous processes   94

3   Finishing methods   99
     PHILIP BALL
     3.1   Introduction   99
     3.2   Cutting   99
     3.3   Surface finishing   101
     3.4   Joining   102

Part III Structural design of polymer composites   105

4   Methodology and management of a design project   107
     ANTHONY MARCHANT

4.1	Introduction	107
4.2	Design considerations	107
4.3	The need for design management	111
4.4	The design process	111
4.5	Design potential	114
4.6	References	115

5   Procedures for designing materials   116
LEONARD HOLLAWAY
5.1	The principles of design analysis	116
5.2	Requirements of materials' design	116
5.3	Characteristics of materials	122
5.4	General laminates	127
5.5	Sandwich construction materials	131
5.6	Concluding remarks	134
5.7	References	135

6   Structural component design techniques   136
ALASTAIR JOHNSON
6.1	Introduction	136
6.2	Design of composite panels	140
6.3	Design of thin-walled beam structures	148
6.4	Design of sandwich structures	163
6.5	Design of thin-walled vessels	171
6.6	Concluding remarks	177
6.7	References	179

7   Limit state design method   181
PETER HEAD
7.1	Introduction	181
7.2	General principles of limit state design	181
7.3	Characteristics of FRP materials	186
7.4	Application of partial coefficient method	189
7.5	Example	193
7.6	References	196

Part IV Joints   199

8   Bonded and mechanically fastened joints   201
FRANK L. MATTHEWS
8.1	Introduction	201
8.2	Joint types	202
8.3	Mechanism of load transfer	202
8.4	Bonded joints	203
8.5	Mechanically fastened joints	223

Contents   vii

8.6   Bolted–bonded joints   240
8.7   References   241

Part V Case studies   **243**

9   Design of bottom shell laminate for GFRP vessel   245
     ANTHONY MARCHANT
     9.1   Bottom skin/stiffener design   249
     9.2   Deck sandwich structure   255
     9.3   References   261

10   The walkways at Terminal 2, Heathrow Airport: Designing
     for fire performance   262
     D. BERRY
     10.1   Introduction   262
     10.2   Fire performance   263
     10.3   Theory of fire performance   264
     10.4   Fire tests   266
     10.5   The future   274
     10.6   The European market   274
     10.7   The cone calorimeter   275
     10.8   Designing the Heathrow Terminal 2 walkway panels:
            Class 0   276
     10.9   Fire resistance   276
     10.10 Conclusions   277
     10.11 Bibliography   278

11   Design of a commercial vehicle sideguard system   281
     PETER HEAD
     11.1   Introduction   281
     11.2   Design development   281
     11.3   Structural analysis   284
     11.4   Design of pultruded beams using the limit state
            design method   285
     11.5   Manufacturing specifications, tooling and production 288
     11.6   Proof testing of the system   289
     11.7   In-service performance   290

12   Design and production of a composite chassis component in
     the automotive industry   291
     G. F. SMITH
     12.1   Introduction   291
     12.2   Development strategy   292
     12.3   Component selection and feasibility   292
     12.4   Component design   296

12.5	Design analysis	296
12.6	Numerically controlled tool production	297
12.7	RIM manufacture	298
12.8	RIM	299
12.9	Process conclusions	300
12.10	Prototype and further work	301
12.11	Design composite storage and access using knowledge based systems	301
12.12	References	302

13 The use of pultruded profiles in a structural application   303
J. A. QUINN

13.1	Introduction	303
13.2	An example of selective design	305
13.3	Bibliography	308

14 GFRP minesweeper: pre-production test structure with box core sandwich construction   309
A. E. DAVEY

14.1	History	309
14.2	Description	309
14.3	Design requirements	309
14.4	Design approach	310
14.5	Structural analysis	312
14.6	Shock loads	312
14.7	Manufacturing challenges	322
14.8	Test results	324

15 Shield for 4.5 inch Mark 8 gun manufactured by Vickers and used on Type 42 destroyers   325
A. E. DAVEY

15.1	Reason for composite materials	325
15.2	Design requirements	325
15.3	Design approach	325
15.4	Structural analysis	326
15.5	Tests	326
15.6	Manufacturing and challenges	326
15.7	Quality assurance for structural reliability	327
15.8	Service experience	327
15.9	References	328

Index   334

Preface

Over the last two decades there has been a growing awareness amongst structural engineers, and not only those working in the aerospace industries, of the importance and use of composite materials. Their utilisation has grown from the intermediate technology systems, which employ the labour intensive hand lay fabrication process, to the high technology automotive manufacturing methods of the pultrusion, filament winding and hot press techniques. The former process was used to make load bearing and semi-load bearing infill panels, mainly isotropic in nature, which became popular in the early seventies and is still utilised. The latter production techniques are used to manufacture high technology composite structures, which are anisotropic in nature; in order to take advantage of the high strength and stiffness, fibres are placed in the most advantageous positions and directions in the structure. To use these composites effectively and efficiently requires detailed analysis and design.

To undertake analysis and design of composite materials may seem a daunting task to many engineers who have training and experience in the conventional structural materials. It is the purpose of this handbook to dispel the mystery of composites by introducing polymers and fibres, to discuss the fabrication techniques in tabular form and by simple equations to give the rudiments of the design of the material, of structural units and of bonded and bolted joint systems.

With new materials, it is imperative to understand how they will behave over their accepted lifetime under load and in the natural environment; consequently, the end of this handbook contains a number of case studies.

The aims of this handbook, therefore, are to introduce the engineer to fibres and polymer matrices which are the components of the polymer composites for structural engineering. In addition, the book will provide a simple guide, in tabular form, to the principal fabrication techniques and simple design formulae and methods for structural composite systems and connections; references for further reading have been given. The chapters do not give an exhaustive picture but it is hoped that they will introduce the design aspects of composites in a clear way, leading to further more advanced investigations in the design techniques.

The handbook is intended for practising structural engineers who need to acquaint themselves with the principles and the structural design techniques of polymer composite materials; third year undergraduate and postgraduate structural students will also find it useful. It is hoped that the approach given in the text will help designers to overcome any initial reluctance to use the material.

Finally, I would like to offer my sincere thanks to the members of the working party whose inspiration, enthusiasm and suggestions have made this handbook possible. There is one member in particular to whom I am indebted: the late Mr Roy Templemen of Maunsell Structural Plastics who was the first chairman of the working party and who was to have been the editor of the Handbook. It is greatly to be regretted that Roy's death in 1991 prevented him from seeing the task through to its fulfilment.

The handbook was conceived by the Composites Group Advanced Composites Section of the British Plastics Federation London and their help and encouragement to enable this book to be published are gratefully acknowledged.

The members of the Composites Group Working Party are:

P. Ball   Armfibre Ltd, Bedfordshire, UK.
L. Hollaway   University of Surrey, Surrey, UK.
A. Johnson   D.L.R. Institute of Structures and Design, Stuttgart, Germany.
A. Marchant   Associated and Marine Technology, Hampshire, UK.
F. Matthews   Imperial College of Science, Technology and Medicine, London, UK.
R. Templeman   Maunsell Structural Plastics Ltd, Beckenham, Kent, UK.

Leonard Hollaway

List of contributors

PHILIP BALL, BSc (Eng), FIM, Operations Director, Armfibre Ltd, Bedfordshire, UK.

D. BERRY, BSc, Formerly Head of Building Services, SGS UK Ltd, Redhill, Surrey, UK.

A.E. DAVEY, MSc, Engineering Manager, Meggitt Composites, Dudley, West Midlands, UK.

PETER HEAD, BSc, ACGI, CEng, MICE, MIStructE, Director, Maunsell Structural Plastics, Maunsell House, Beckenham, Kent, UK.

LEONARD HOLLAWAY, MSc(Eng), PhD, CEng, FICE, MIStructE, Professor of Composite Structures, Department of Civil Engineering, University of Surrey, Guildford, UK.

ALASTAIR JOHNSON, BSc, PhD, CEng, FIM, Research Engineer, Demonstration Centre for Composite Materials, DLR – Institute for Structures and Design, Pfaffenwaldring 38-40, D-70569 Stuttgart S, Germany.

ANTHONY MARCHANT, CEng, FIMechE, FICE, FRINA, MRAeS, Director, Associated and Marine Technology, Romsey, Hampshire, UK.

FRANK L. MATTHEWS, BSc(Eng), ACGI, CEng, FRAeS, FIM, Director, Centre for Composite Materials, and Senior Lecturer, Department of Aeronautics, Imperial College of Science, Technology and Medicine, London, UK.

LESLIE S. NORWOOD, BTech (Hons), PhD, Head of Applied Technology Group, Scott Bader Research and Development Department, Scott Bader, Northamptonshire, UK.

J.A. QUINN, CEng, MIMfgE, James Quinn Associates Ltd, Consulting Engineers in Composite Materials, Liverpool, UK.

G.F. SMITH, PhD, Principal Research Fellow, Rover Advanced Technology Centre, University of Warwick, UK.

Part I

Introduction

1 Fibre reinforced polymers

1.1 History of fibre reinforced polymer materials

Although molten glass fibres were commonplace over 3000 years ago, their potential as reinforcing materials was not recognised until the introduction of plastics during this century. In fact, apart from natural composite technology (e.g. in the form of wood), today's composite industry can only be said to be born with the introduction of the first thermosetting plastics in 1909: phenolics. However, it was not until the 1940s that the growth of the structural composite industry really began and the *'Synthetic Composite Materials Age'* was upon us.

The first range of thermosetting resins to make their mark in the modern composite age were unsaturated polyester resins, which were introduced in the UK in 1946 by 'Scott Bader' from 'Marco' in the USA. These hot-cure resins were not user-friendly, but did make possible the production of low pressure mouldings of great strength, using relatively simple, cheap production processes. Development was rapid and by 1947 the first British cold-curing polyester resin had been introduced, followed, in 1949, by the first non-air-inhibited resins, paving the way for cheap, simple, open mould production methods.

Up to this time the main commercial use of fibre reinforced plastics (FRPs) was for the construction of aircraft radomes in the USA, from about 1942. Hence, at the same time as developments were being made in polyester resin, so attention was switching to reinforcements. The early radomes were manufactured using glass fibre reinforcement consisting of fine woven glass fibre cloth made from an exceptionally fine glass filament. In the specialised area of aircraft application, cost was of secondary importance, but as the use of composites spread, commercial pressure resulted in the development of a cheaper, coarser glass filament with good mechanical properties, and from 1947 work was devoted to making cheaper forms of reinforcement other than woven cloth – chopped strand mat (CSM) appeared for the first time. Hence, the right reinforcement and the right resin at the right price saw the first FRP translucent sheet production in 1949 followed rapidly in the early 1950s by developments in FRP boat hulls, FRP car bodies and FRP lorry cabs.

As more specialised chemical resistant, heat resistant and fire retardant unsaturated polyester resins developed, so too did the FRP chemical pipe and tank industry and the FRP cladding panel industry. Such developments have seen the world production of reinforced unsaturated polyester resins rise rapidly from 50 000 tonnes in 1955 to 500 000 tonnes by 1968, with current estimates in excess of 2 500 000 tonnes.

Of course, unsaturated polyester resins are just a part of the overall family of thermosetting resin and whereas they were first developed in the USA, epoxy resins originated in Europe in 1937. It was not until around 1950 that commercial production of epoxy resins became possible as a result of the availability of epichlorhydrin in bulk. The use of fibre reinforced epoxide resins has been inextricably linked to the aircraft industry, not least because supply can be in the form of preimpregnated (prepreg) material. Their slow penetration into other reinforced plastic market areas can, no doubt, be attributed to cost and the need for heat assisted cure, although more recent developments have seen the introduction of 'cold cure' epoxy resins.

There are, of course, many other important thermosetting resins available today, which can be used to form special fibre reinforced components, e.g. polyimides, phenolics, vinyl esters, furanes, silicones, polyurethanes and urethane acrylates. It is interesting to note the resurgence of the first thermosetting resins – phenolics – as important materials in modern low smoke, fire resistant FRP applications.

The foundation for the development of the other major class of plastics – thermoplastics – was laid down before that of thermosets, and whilst thermoplastics development has been rapid (and modern lifestyles would be impossible without them), their use in reinforced applications has been restricted and slow growing. The main reasons for this include processing difficulties, low thermal resistance and low mechanical performance. When only reinforced with short fibres, which become further degraded during the injection moulding process, thermoplastics offer little competition to reinforced thermosets in structural applications. However, the recently introduced polyether-ether ketone (PEEK), although expensive, is gaining acceptance in high performance applications, where light weight, temperature resistance and high mechanical properties are essential.

As with resin development, the pace of fibre development has accelerated since the 1950s with the introduction of E glass, R glass, S glass and special acid- and alkali-resistant glass. In addition, textile processing technology has enabled glass rovings to be processed into an enormous variety of fabrics, which have been brought to the market-place in the form of CSM, woven roving (WR), stitched cross-plied rovings, combination fabrics (CSM–WR), unidirectional tapes and fabrics, woven cloth and multidirectional fabrics.

Glass fibre has been the major reinforcement for the FRP industry,

but the drive for lighter, stronger, stiffer structures has seen the introduction of carbon and polyaramid fibres.

Carbon fibre is not a new material, being first used about a hundred
years ago as filaments for electric lamps. This early fibre was relatively
weak and of little use as a reinforcement. By comparison, today's carbon
fibres are leagues apart from their predecessors with exceptionally
high strength and stiffness. However, since the development of this
second generation carbon fibre in 1963, their growth in FRP has been
limited to specialised applications where cost is relatively unimportant,
e.g. aircraft and sports goods.

Polyaramid fibres were first developed in 1965 and although not as
strong or stiff as carbon fibres, they offer considerable performance
advantages over glass fibres. They are lighter than carbon fibre, less
expensive, and exhibit exceptional impact resistance. In the FRP
industry polyaramid fibres have found acceptance in gas pressure bottle
construction, aerospace applications, sports goods and marine applications.

As a result of the introduction of carbon and polyaramid fibres
weaving with glass fibres has led to a proliferation of 'hybrid' fabrics
where the optimum properties of the different fibres can be utilised at
optimum cost.

FRP has come a long way in less than half a century and truly
deserves the description 'composite', i.e. the combination of two or
more materials on a macroscopic scale to form a useful material, often
exhibiting characteristics that none of the components exhibit independently.

1.2 Resins

On their own, bundles of parallel fibres are of little use in a load-bearing
structure. They may have structural integrity in tension but, unless they
can be joined, their structural potential cannot be harnessed. Similarly,
bundles of fibres are little more than useless in shear and compression.
Without a means of distributing load across a series of fibre bundles,
the material is of no use for structural applications other than rope,
which works in cable tension.

By 'glueing' bundles of fibres together with resin matrices, materials
are made where the strong, stiff fibres are able to carry most of the
stress whilst the matrix distributes the external load to all the fibres as
well as affording protection and preventing fibre buckling under compressive forces. The most critical region for the load transfer process is,
of course, the fibre–resin interface; this interface is often poorly
defined, particularly so in the case of glass fibre where an aqueous
mixture of coatings is applied during the fibre manufacturing process.

However, since size technology has become part of the fibre manufacturer's 'art', this will be considered in more detail in Section 1.4.

1.2.1 Unsaturated Polyester Resins

1.2.1.1 Structure

The most frequently used resins for the manufacture of FRP are unsaturated polyester resins, usually simply referred to as polyesters. They are syrups consisting of polymer chains dissolved in a reactive organic solvent (monomer). Addition of a suitable catalyst and accelerator causes the syrup to undergo a chemical reaction in the cold state and, without pressure, to form a solid, three-dimensional structure. This can be depicted simply as shown in Fig. 1.1 consisting of polyester polymer chains joined to reactive solvent polymer chains. In this example, styrene is the reactive organic solvent and the reactive solvent polymer chains are polystyrene.

Since the polyester to styrene reaction is an order of magnitude faster than the styrene to styrene reaction, the polystyrene links

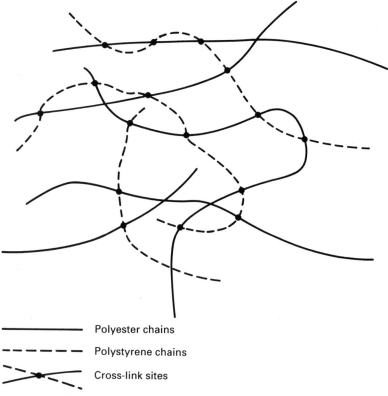

—————— Polyester chains

– – – – – – Polystyrene chains

Cross-link sites

1.1 The structure of cured polyester resin.

between the polyester chains tend to be only one styrene molecule in length.

The characteristics of polyester resins can be varied by altering the ratios of their chemical building blocks during the production process. The main starting materials are dibasic organic acids (both saturated and unsaturated) and dihydric alcohols (glycols). By the elimination of water between the acids and the glycols, ester linkages are formed, resulting in a long chain molecule built up from alternate acid and glycol units (as shown in Fig. 1.2). Careful regulation of the ratio of saturated to unsaturated dibasic acid results in the incorporation of cross-link sites (unsaturation or carbon–carbon double bonds) at regular intervals along the chain. Varying the density of the unsaturated sites in the polymer chain has fundamental effects on the cured structure of the resin and its final properties, although the full impact of high unsaturation is only achieved with high temperature post-cure, as will be seen later, when the cross-linking reaction goes to completion.

1.2 Polyester polymer chain.

The principal acids and glycols used in the manufacture of polyester resins are given in Table 1.1. Whereas the level of unsaturated acid plays an important role in determining the cross-link density of a polyester resin when reacted with monomers (such as styrene, methyl methacrylate, vinyl toluene and alpha-methyl styrene), other properties such as chemical resistance and fire resistance will be influenced by the saturated acid and glycol used. Polyesters are therefore commonly classified according to the material used in their manufacture (orthophthalic, isophthalic, iso-NPG, bisphenol). Of course, cross-link density has a marked influence on properties and the effect is most noticeable when comparing the same resin in the cold-cured and fully post-cured state.

Other factors important to resin properties are processing (which largely determines the polymer chain length), monomer content and filler addition, e.g. for fire retardant properties.

1.2.1.2 Curing
Polyester resins are cured via a free radical co-polymerisation reaction, which can be initiated at room temperature with organic peroxides, by

Table 1.1 Acids and glycols used in the manufacture of polyester resins

Unsaturated acids	Saturated acids	Glycols
Maleic	Orthophthalic	Propylene
Fumaric	Isophthalic	Ethylene
Maleic in the form of maleic anhydride	Terephthalic	Di-ethylene
	Hexachloro-endo-methylene-tetra-hydrophthalic ('HET' acid)	Neopentyl (NPG)
	Orthophthalic in the form of phthalic anhydride	Dihydroxy dipropoxy bisphenol 'A' Dibromoneopentyl

heat or by radiation, such as UV and visible light. It is the ability to cold-cure polyester resins from liquid that makes them more suitable than any other resin system for the manufacture of very large structures.

The catalyst systems consist of organic peroxides (initiators) activated by accelerators or promoters. Peroxides used, although referred to as 'catalysts', are not so in the generally accepted chemical definition of a catalyst because they do not remain unchanged during the free-radical reaction. In fact, the true chemical catalyst is the accelerator, which promotes the breakdown of the peroxide to initiate the reaction.

The accelerator promotes the initiator (peroxide) decomposition by rupturing the peroxide link to form two free radicals. Each of these retains one electron from the broken electron pair and is a highly reactive species. A 'chain' reaction then occurs as peroxide free radicals capture electrons from the vinyl monomer, creating another free radical which reacts with the unsaturation in the polyester chain and so on, as shown below:

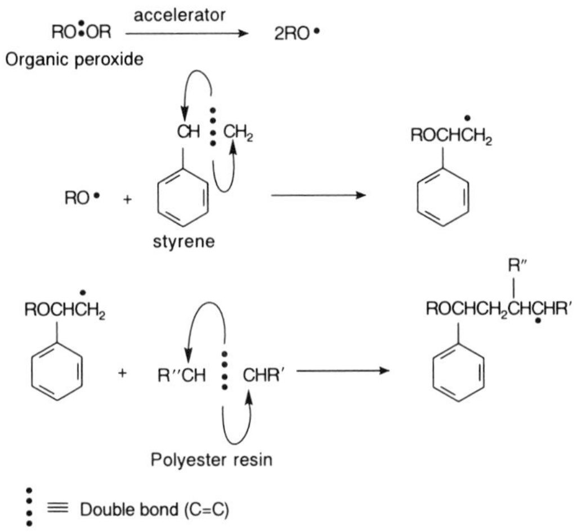

Peroxides are broken down with promoters, such as cobalt octoate, and tertiary aromatic amines, such as dimethylaniline and diethylaniline. The most common peroxides used for curing polyester resins are methyl ethyl ketone peroxide (MEKP), acetyl acetone peroxide, cyclohexanone peroxide and benzoyl peroxide.

The cure of a polyester resin begins as soon as initiator is added. The speed of the reaction depends upon temperature and the resin and catalyst reactivity. Without the use of the accelerator, heat, or radiation, the catalysed resin will have a pot life of hours or even days. It is, therefore, essential to have sufficient quantities of initiator and accelerator present to ensure that adequate cure occurs. There is no recovery from initial undercure, even with high temperature post-cure. Curing takes place in several stages:

1 The induction period, where nothing seems to happen. However, free radicals are being formed and immediately reacting with inhibitor, which is used to give resins adequate shelf life.
2 Gelation: the cross-linked network structure begins to form.
3 Exotherm builds up.
4 Final hardening (this can take days or weeks).

Establishing complete cure has been the subject of much study, but there is no simple answer. A complete picture can be obtained by a comprehensive examination of a wide variety of properties such as solvent extraction, residual styrene, hardness measurement, static and dynamic mechanical tests, infra-red absorption, chemical resistance and electrical tests.

1.2.1.3 Cast Resin Properties

Typical short term cast properties of various types of fully cured polyester resin are shown in Table 1.2 after post-cure at 80 °C. Some general relationships between composition and properties are apparent:

1 More flexible resins (greater elongation) have lower unsaturated : saturated acid ratios.
2 Higher unsaturated : saturated acid ratios (greater cross-link density) result in higher heat deflection temperature (HDT).
3 Longer chain glycols (e.g. diethylene glycol) result in higher elongation, but lower HDT values.
4 Isophthalic acid can be processed to longer chain length compared with orthophthalic acid, resulting in greater strength, greater elongation and improved water resistance.
5 Fire retardant fillers increase the modulus and specific gravity, but reduce elongation.

In many applications resins are cold-cured and the maximum properties given in Table 1.2 are not achieved. Some typical 28-day room temperature cured properties are shown in Table 1.3 for a selec-

Table 1.2 Typical properties of fully cured cast polyester resin

Resin type	Tensile Str. (MPa)	Tensile Mod. (GPa)	Tensile El. (%)	Flexural Str. (MPa)	Flexural Mod. (GPa)	Compression Str. (MPa)	Compression Mod. (GPa)	HDT (°C)	24 h H$_2$O Abs. (mg)	Barcol hardness	Specific gravity
High reactivity orthophthalic	54	3.6	2.0	136	3.9	130	3.7	110	30	48	1.21
Medium reactivity orthophthalic	68	3.6	2.4	113	3.7	134	3.7	75	20	46	1.23
Low reactivity orthophthalic	60	3.8	2.0	127	4.2	137	4.4	65	16	45	1.22
High reactivity isophthalic	55	3.3	2.0	101	3.8	122	3.3	125	30	48	—
Medium reactivity isophthalic	66	3.5	2.4	122	4.0	135	3.7	95	21	46	1.23
Low reactivity isophthalic	80	3.3	4.5	135	3.3	125	3.5	75	19	43	1.21
Medium reactivity iso-NPG	60	3.2	2.5	92	2.6	110	2.2	120	19	40	1.14
Low reactivity iso-NPG	70	3.2	3.7	122	3.4	125	3.6	95	14	43	1.16
Bisphenol	67	2.9	3.4	135	3.4	112	3.0	120	18	37	1.14
Medium reactivity clear HET acid	60	2.8	3.5	110	3.1	106	2.7	67	31	38	1.26
Low reactivity clear HET acid	50	3.5	1.5	90	3.9	129	3.8	70	12	45	1.25
Low reactivity filled HET acid	50	5.1	0.8	76	4.9	101	4.4	59	9	45	1.30

Medium reactivity filled FR orthophthalic	52	5.3	1.1	76	5.4	94	3.7	59	13	46	1.56
Low reactivity filled FR orthophthalic	56	4.0	1.6	114	4.1	134	4.0	65	14	47	1.32

Str. = strength; Mod. = modulus; El. = elongation; Abs. = absorption; **FR** = fire retardant.

Table 1.3 Typical properties of low temperature cast polyester resin (16 hours at 40 °C or 28 days at room temperature)

	Property					
Resin type	Tensile strength (MPa)	Tensile modulus (GPa)	Tensile elongation (%)	HDT (°C)	24 hours water absorption (mg)	Barcol hardness
Medium reactivity orthophthalic	64	3.0	4.5	57	18	42
Low reactivity orthophthalic	60	3.2	3.0	55	15	43
Medium reactivity isophthalic	55	2.7	4.8	59	20	35
Low reactivity isophthalic	57	2.6	7.5	54	18	35
Medium reactivity iso-NPG	48	2.6	4.7	62	16	32
Low reactivity clear HET acid	52	3.2	2.0	58	10	40

tion of resins. In general, the elongation is higher and the HDT lower for non-post-cured cast polyester resin.

While the resin's physical and mechanical properties are seen to be quite quantifiable, chemical resistant properties are less easily quantified. In general, the resistance increases going from orthophthalics to isophthalics to iso-NPGs to bisphenols. Again, the level of post-cure directly influences the level of chemical resistance, especially when the resin is to be exposed to elevated temperatures during service. Figure 1.3 emphasises how the level of post-cure affects cast resin properties by comparing the HDT of a variety of resins after post-cure at various temperatures. It is noticeable, from Table 1.3 and Fig. 1.3, that low temperature post-cure, e.g. 16 hours at 40 °C (= 28 days at room temperature) results in little difference between the HDTs of both low and high cross-link density resins. Hence, to exploit the optimum temperature resistance of some resins, elevated temperature post-cures, at least at 100 °C, are essential.

Much criticism is made of the toughness of polyester resins, but systems for increasing flexibility are available, which are totally miscible with polyester resins, and result in considerable improvement in flexibility whilst still giving adequate retention of other important

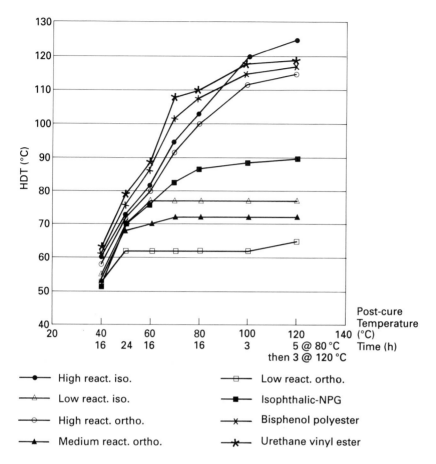

1.3 Variation of HDT of cast polyester resin with post-cure temperature.

properties. These materials are urethane-acrylates in styrene monomer. Typical properties of polyester resin made more flexible with urethane-acrylate resin are shown in Table 1.4.

Other typical properties of cast polyester resin (e.g. thermal and electrical) are given in Table 1.5. As with the properties discussed earlier, these will be greatly influenced by the presence of reinforcing fibres and fillers in reinforced composites.

1.2.2 Vinyl Ester Resins

1.2.2.1 Structure
Vinyl ester resins are best considered as an extension of the polyester resin range in that they are supplied in a reactive monomer, usually

Table 1.4 Typical properties of cast flexibilised (with urethane acrylate) polyester resin (low reactivity isophthalic)

Property	Proportion of flexibiliser (%)	Fully cured							Room temperature (or 16 hours at 40 °C) cured						
		0	10	20	30	50	70	100	0	10	20	30	50	70	100
Tensile strength (MPa)		70	55	50	44	30	27	23	65	55	50	40	38	26	23
Tensile modulus (GPa)		3.3	2.7	2.1	1.6	0.8	0.6	0.5	3.3	2.7	3.2	2.3	2.3	0.7	0.4
Elongation (%)		4.5	6.5	7.5	13	17	50	110	5.5	6	7	14	17	48	115
Notched izod impact strength (J/m)		*	21	25	26	70	105	142	29	33	38	42	80	**	**
Charpy impact strength (kJ/m²)		9	15	16	21	34	58	**	11	8	23	25	26	**	**
Drop weight impact strength (kg cm)		3	3	5.5	8	14	35	>200	3.5	3.5	5	7.5	9	45	>200
Water absorption (7 days) (mg)		46	47	52	54	58	63	69	47	43	50	53	54	60	62
Heat deflection temperature (°C)		72	64	62	55	44	37	25	55	55	52	53	50	39	20

* Unnotchable.
** No failure.

Table 1.5 Electrical and thermal properties of fully cured cast polyester resin

Specific heat (kJ/kg K)	2.3
Thermal conductivity (W/mK)	0.2
Coefficient of linear expansion (10^{-6}/°C)	100
1000 Hz dielectric constant	2.75
50 Hz power factor	0.008
1000 Hz power factor	0.004
5 MHz power factor	0.019
50 Hz permittivity	3.7
5 MHz permittivity	3.2
Voltage breakdown (0.2 mm sample) (kV/mm)	22
Volume resistivity (T Ω m)	1

styrene, and cure via a free radical reaction initiated in the same way as the cure mechanism for polyester resins. The typical chemical structure of vinyl ester resin is shown below:

The main backbone is the bisphenol 'A' epoxy structure, but this is different from the structure of a conventional bisphenol polyester resin with unsaturation only at the end of the chains and not in the repeat unit. Also, vinyl resins contain fewer ester linkages than conventional polyesters and have, hence, improved chemical resistance, whilst the terminal double bonds result in a tougher, more resilient resin structure.

1.2.2.2 Curing
Vinyl ester resins are processed in a manner similar to polyester resins, using peroxide catalysts and cobalt accelerators, often boosted with the addition of dimethylaniline. Hot-cure systems based on benzoyl peroxide or tertiary-butyl peroxybenzoate can also be used.

1.2.2.3 Cast Resin Properties
Vinyl ester resins have similar properties to a number of polyester resins, but as with polyester resins, the maximum temperature resistance is only fully achievable when high temperature post-cure is used (Fig. 1.3). At low temperature or room temperature post-cure, they exhibit little or no benefit over polyester resins. Table 1.6 shows typical properties of both room temperature and fully cured vinyl ester resins.

Table 1.6 Typical properties of cast vinyl ester resin

Resin type / Property	General purpose vinyl ester fully cured	Urethane vinyl ester	
		Fully cured	Room temperature cured
Tensile strength (MPa)	70–80	65	55
Tensile modulus (GPa)	3.3	3.2	2.8
Elongation (%)	5–6	4.5	6.5
Flexural strength (MPa)	130–140	—	—
HDT (°C)	95–125	125	60
Charpy impact (kJ/m^2)	—	16	—
Notched izod impact (J/m)	—	50	—
Water absorption (mg)	—	16	15
Barcol hardness	—	40	30

As emphasised for polyester resins, a degree of elevated temperature post-cure is essential if the maximum performance and optimum chemical resistance of vinyl ester resins are to be achieved.

1.2.3 Epoxy Resins

1.2.3.1. Structure

Epoxy resins are generally manufactured by reacting epichlorohydrin with bisphenol 'A'. They are viscous liquids in their thermoplastic state and differ significantly from polyester and vinyl ester resin in that they contain no volatile monomeric component. Different resins are formed by varying the proportions of epichlorohydrin and bisphenol 'A': as the proportion of epichlorohydrin is reduced, so the molecular weight of the resin is increased. Resins produced in this way consist of mixtures of different molecular weight species whose distribution governs viscosity/melting point characteristics:

As a consequence of their high viscosity, epoxy resins are frequently processed at elevated temperatures (50–100 °C) or dissolved in an inert solvent to reduce viscosity to a level where lamination at room temperature becomes possible. However, this can introduce problems in the presence of low temperature curing hardeners, since the resin may cure before all the solvent has evaporated.

Other epoxy resins
Glycidyl ester resins have the following structure:

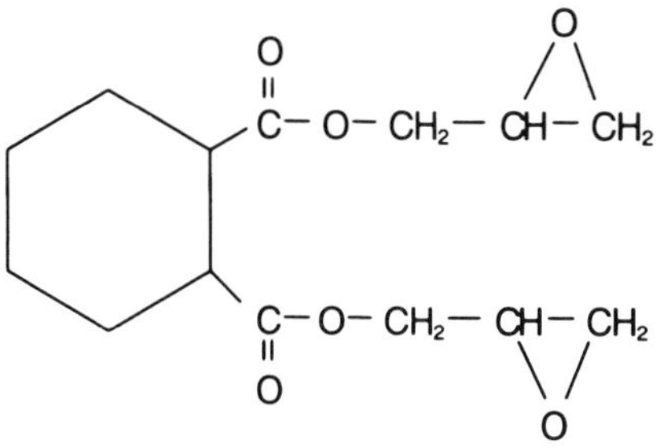

These have viscosities and reactivities ideally suited to vacuum impregnation, laminating and casting applications.

Glycidyl ethers of novolac resins are made by reacting novolac resins, produced by reacting phenol and formaldehyde together:

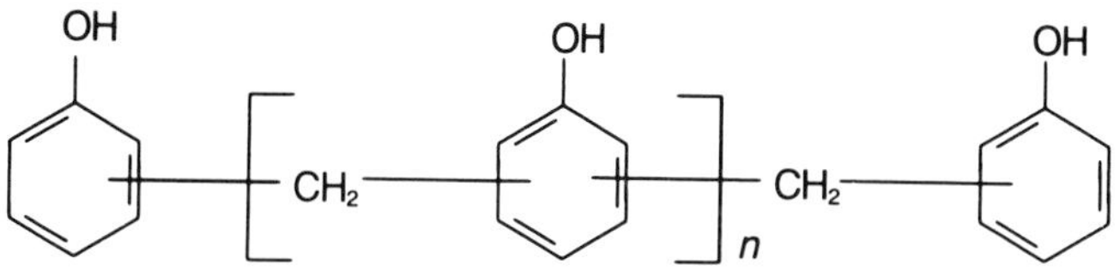

with epichlorohydrin to form glycidyl ether novolacs.

Brominated resins include the diglycidyl ether of tetrabromobisphenol 'A' and glycidyl amine resins.

1.2.3.2 Curing

Epoxy resins are cured by means of a curing agent, often referred to as catalysts, hardeners or activators, and although the terms are often used indiscriminately, there are differences between them. Some curing agents work via catalytic action, whereas others react with the resin and become absorbed into the resin chain. During curing, epoxy resins can undergo three basic reactions:

1 Epoxy groups are rearranged and form direct linkages between themselves.
2 Aromatic and aliphatic hydroxyl groups react with epoxy groups.

3 Cross-linking occurs with the curing agent via various radical groups.

Amine curing agents may be primary or secondary, aliphatic, alicyclic or aromatic. The proportion of epoxy resin to amine hardener is modified to achieve optimum cured properties based on the following reaction:

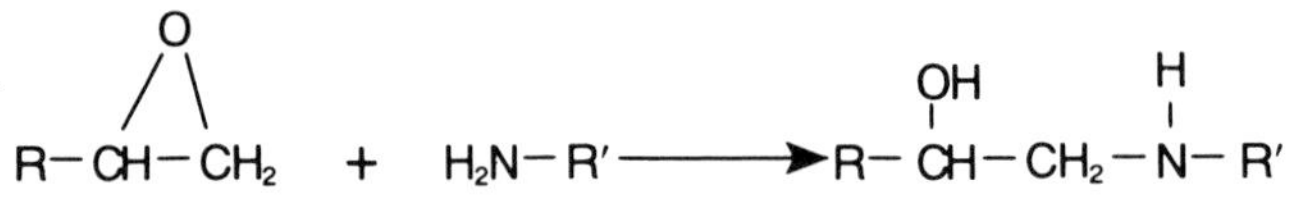

The hydroxyl group can react further and the secondary amine formed reacts with another epoxy group:

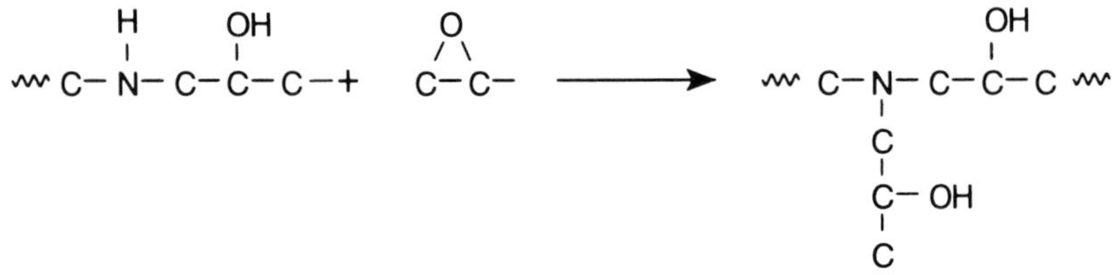

Since both the epoxy resin and the amine are polyfunctional, a highly cross-linked structure is created. The epoxy-hydroxyl reaction does not generally occur in this type of curing.

Anhydrides react by combining with both the epoxy groups and the hydroxyl groups present in most epoxy resins. The reaction is initiated by the formation of the half ester through the anhydride joining the hydroxyl group. The remaining acid group then reacts with an epoxy group. Self-polymerisation of the epoxy group through the epoxy-hydroxyl reaction may also occur.

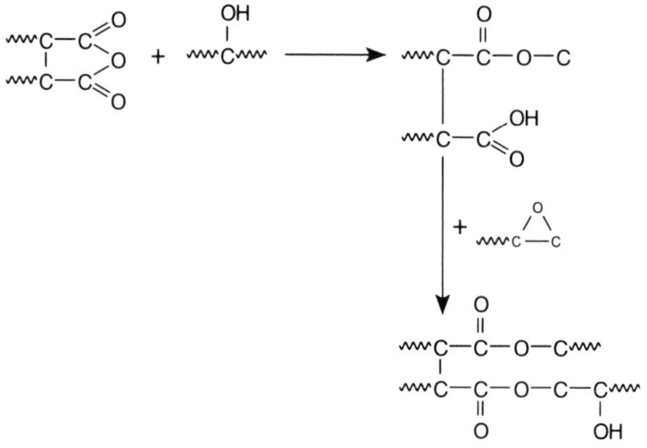

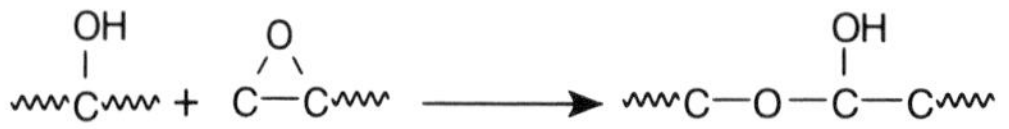

Table 1.7 Typical properties of fully cured cast epoxy resin

Curing agent	Property						
	Tensile strength (MPa)	Tensile modulus (GPa)	Elongation (%)	HDT (°C)	Flexural strength (MPa)	Permittivity at 1 kHz	Loss tangent at 1 kHz
Diethylene-triamine	70	3.4	5.3	95	—	4.2	0.040
Isophorone-diamine	82	2.8	—	150	116	3.9	0.016
Cyclohexyl-propylene-diamine	65	2.7	—	95	90	4.1	0.016
Diaminodiphenyl-methane	80	2.8	5.2	155	117	4.1	0.006
Diaminodiphenyl-sulphone	78	3.1	6.0	193	—	4.0	0.017
Phthalic anhydride	82	3.0	4.1	110	—	3.6	0.002
Tetrahydrophthalic anhydride	81	2.7	4.5	120	—	3.4	0.003
Chlorendic-anhydride	82	3.2	2.6	186	—	3.3	0.006
Pyromellitic dianhydride	22	2.8	—	290	—	4.0	0.002

Thermal properties

Thermal conductivity (W/m °C)	0.21
Coefficient of expansion (10^{-6}/°C)	80–120

The favoured reaction is dependent upon the curing temperature, accelerator used, etc.

1.2.3.3 Cast Resin Properties

Typical properties of cast epoxy resin are shown in Table 1.7 for different curing systems. It is very important to note the need, generally, to cure epoxy resins at an elevated temperature to achieve any level of acceptable cure. Although 'cold-cure' epoxy resins are available, they produce better properties when cured above 40 °C and preferably at 60 °C.

1.2.4 Other Thermosetting Resin Systems

There are other resins besides polyesters, vinyl esters and epoxy resins which are used for laminating and moulding applications. These include furane resins, polyimide resins, silicone resins, phenolics, melamine and urea–formaldehyde resins. Of these, the oldest known thermosetting resins – phenolics – are finding a new lease of life because of their low smoke and fire retardant characteristics.

1.2.4.1 Phenolic (phenol-formaldehyde) Resins

Phenolic resins are produced via a condensation reaction by combining phenol and formaldehyde in the presence of a catalyst.

The acid catalyst used in the cure of phenolic resins is a major drawback to their use from a handling point of view. Apart from their advantageous fire resistant performance, phenolic resins offer few advantages over polyester resins, being brittle, limited in colour and difficult to process.

1.2.4.2 Silicone Resins

Silicone resins are the most heat resistant resins used in the manufacture of fibre reinforced plastic. They exhibit good mechanical, electrical and thermal properties. Silicone resins are cross-linked by heating in the presence of a catalyst such as cobalt naphthenate, zinc octoate or amines such as triethanolamine.

Silicone resins are often used in prepreg manufacture.

1.2.4.3 Polyimide Resins

Polyimide resins are some of the best thermally stable organic resins currently known. Moulding is carried out at around 300 °C with a post-cure at 400 °C to obtain full thermal stability. The elimination of water during cure is another added complication.

1.2.4.4 Furane Resins

Furane resins are primarily used for chemical resistant applications as they have probably the best chemical resistance of any thermosetting resin in non-oxidising conditions. They have excellent solvent resistance, unlike many of the other commonly used resins in chemical plant applications.

Like phenolic resins, furane resins require acidic catalysts to cure, hence presenting considerable fabrication problems.

1.2.5 Thermoplastic Resins

There is growing interest in the use of thermoplastics in long fibre reinforced applications because of their excellent toughness, resilience and corrosion resistance. These straight chain polymers, such as polycarbonates, nylons, polyethylenes, polypropylenes and polystyrenes do, however, have fundamental disadvantages compared with thermosetting resins, in that they have to be moulded either at elevated temperature, i.e. they are softened, or above their melting point. They are very viscous under molten conditions and in the finished state have lower levels of heat and solvent resistance.

Much of the literature on thermoplastics will indicate superior temperature resistance compared with thermosetting resins, especially comparisons of HDT, but such data are misleading because the measurement of HDT for thermoplastics is allowed under a load of just 0.45 MPa, whereas thermosetting resins have to be subjected to a load of 1.8 MPa to measure the same property. If thermoplastics were subjected to 1.8 MPa under the HDT test conditions, it is unlikely that many would exhibit HDTs much above room temperature.

Many 'reinforced' thermoplastics are manufactured by injection mouldings where the fibre lengths are very short – often no more than 30 μm – because of mechanical damage. Hence, they are better considered as filled systems for non-structural, low temperature applications.

1.3 Additives to Resins

Rarely are base resins supplied without the addition of particulate materials, which are generally present to improve their performance. For example:

1   Decorative finish – colour by use of pigment pastes.
2   Handling – thixotropic agents (glass microspheres).
3   Fire retardancy – halogenated material, antimony trioxide, alumina trihydrate.
4   Light stabilisation – UV absorbers.
5   Air inhibition reduction – surface migrating material.
6   Air release promotion.
7   Improved modulus.

Of course, with benefits there can often be penalties, and this is so with the addition of fillers to resins where a reduction in flexibility and opacity are almost inevitable. However, the benefits often outweigh the disadvantages, and only where cheap fillers are added in higher quantities for cost reduction purposes is there a need to question the benefits of such practices in terms of component performance.

1.3.1 Types of Filler

Fillers can be classified into a number of groups.

1.3.1.1 Mineral Fillers

Many of the fillers used in resin modification are obtained from naturally occurring deposits, such as:

1   *Carbonates*: calcium carbonate is predominantly used in the form of chalk or calcite (limestone).

2 *Magnesium carbonate*: this is used in the form of magnesite or dolomite, which is a combination of both magnesium and calcium carbonates.

1.3.1.2 Silica and Silicates

The most common pure forms of silica used as resin additives are quartz, silica sand and silica (quartz) flour. In addition, silicates such as talc (magnesium silicate), calcium silicate (occurring as the metasilicate wollastonite), kieselguhr (diatomaceous earth) and zircon (zirconium silicate) find frequent applications.

Clays (aluminium silicates) are, of course, frequently used as inexpensive fillers. In particular, china clay (kaolin) can act as a flow control agent and thixotrope, as well as giving high gloss surfaces.

Slate (highly compacted sedimentary clay), *mica, calcium aluminium silicate, vermiculite* and *pumice* are also used with resins.

1.3.1.3 Glass Fillers

1 *Ballotini* (solid glass microspheres) coated with silicone coupling agent, to improve adhesion, increase modulus, chemical and abrasion resistance, whilst reducing water absorption without significantly altering the resins' handling characteristics.
2 *Hollow glass spheres* are used to reduce density and can improve thixotropy.
3 *Glass flakes* are mainly used in coatings, such as gelcoats, to reduce chemical and water permeation and to provide a high wear, abrasion resistant surface.

1.3.1.4 Metal Oxide Fillers

Metal oxide fillers are used for a variety of reasons. Some act to improve thermal characteristics of resins, some as fire retardants, and others are excellent pigments:

1 *Metal oxide pigments* include the oxides of iron, lead, titanium and zinc.
2 *Metal oxide flame retardants*: antimony trioxide and alumina trihydrate are often used to reduce flammability of polyester resins.
3 *Metal oxide thermal additives*: aluminium oxide (alumina) and magnesium oxide are often used to improve the thermal conductivity of resin. In addition, alumina imparts hardness and abrasion resistance whilst magnesium oxide is a useful thickener in moulding compound production (such as sheet and dough moulding compounds, SMC and DMC).

1.3.1.5 Metal Powder Fillers

Powdered metal fillers are often added to resin for purely decorative purposes. However, some metal powders have been used to improve electrical and thermal performance of resins, whilst copper powder is used to reduce marine fouling.

1.3.1.6 Other Additives

1 *Carbon*, in the form of carbon black or graphite, is both a pigment and a way of making cured resin systems electrically conductive.
2 *Fillite*: these glass-hard, inert, lightweight silica spheres can be used up to 70% loadings by weight to reduce weight in the final product.
3 *Silica aerogel* or *fumed silica* is a very important thixotrope, which is effective with both epoxy and polyester resins; its use reduces the problems of drainage when laminating large structures with varying contour.
4 *Pigments*: apart from those already mentioned, many inorganic and organic compounds are used to provide a wide range of colours to surface finishes for laminating materials. The resins are generally supplied pre-coloured, but pigment pastes are also available for use by the fabricator.

Resins, both clear and coloured, can and are used on their own, but by far their most important use is in the manufacture of fibre reinforced structural components for industries as diverse as transport and building, potable water storage and corrosive chemical storage, marine/offshore and sporting goods. For these applications, the use of fibres – which can be in many forms – is essential.

1.4 Reinforcements

Reinforcements are used with resin systems to improve the mechanical properties of cured resin and provide usable components. By far the most important fibre used with polyester and epoxy resins is glass fibre, which is supplied in a variety of forms, e.g. continuous rovings, woven rovings, cloths and random chopped fibre mats.

In recent years high strength carbon fibres and polyaramid fibres have found increasing use in the manufacture of composite materials for all applications, often in the form of hybrid products where the best features of each constituent are utilised to the full. The high strength and stiffness-to-weight ratios of carbon and polyaramid fibres make them particularly attractive for the manufacture of lightweight structural components.

Natural fibres such as jute and sisal have also been used to reinforce thermosetting resins, but their application has been limited by their poor long term environmental performance.

Since glass fibre is the most widely used reinforcement for fibre reinforced plastics, much of this section will be devoted to its production, properties and fabrics.

1.4.1 Glass Fibre Reinforcement

Glass fibre reinforcement is supplied in a variety of forms: A, C, E, R and S. 'A' or alkali glass was at one time the common base material for glass fibre production. Today, this has been superseded by 'E' or electrical grade glass, a very low alkali content borosilicate glass providing good electrical, mechanical and chemical resistant properties. 'C' glass is a special chemical resistant glass used for surface tissue manufacture, whereas R and S glass are high strength glasses supplied in fibre form, mostly for aerospace applications. Typical compositions for some types of glass are shown in Table 1.8.

1.4.1.1 Manufacture

Blends of inorganic material are fused and reacted together in a high temperature melting furnace. At around 1600 °C dissolved gases are removed and homogeneity of the glass composition is obtained. In the direct melt system the refined glass is fed directly to filament, forming platinum 'bushings', but in the marble process the refined glass is made into marbles which are annealed and subsequently remelted in platinum bushings.

In both processes an electrically heated platinum bushing, containing between 100 and 800 drawing points, is used. The molten glass is drawn through the bushing into continuous filaments, at a speed of 50 m/s. The individual filaments are coated with size and other additives as an emulsion before being brought together in strands, which are then wound on to high speed collets. The basic packages of glass strands are usually referred to as 'cakes'. The filament diameters range from 3.25 to 14 μm or more, with 'E' glass filaments typically around 13 μm in diameter. The linear density of glass fibres is measured in 'tex' – grams/kilometre.

Table 1.8 Typical glass compositions

| Component | Composition (%) Type of glass | | | |
	E	C	S	A
Component				
SiO_2	52.4	63.6	64	72.5
Al_2O_3 Fe_2O_3	14.4	4.0	26	1.5
CaO MgO	21.8	16.6	10	12.5
B_2O_3	10.6	6.7	—	—
Na_2O K_2O	0.8	9.1	—	13.5

1.4.1.2 Forming Sizes

Forming size is applied to glass fibre filaments shortly after the drawing process, and performs a number of functions including the application of linking agent to the glass, protection of the glass from interfilament abrasion, and as a processing aid to optimise the fibres' performance in the ultimate end use. Typical forming sizes are applied as emulsions and consist of:

1 A linking or coupling agent, which is typically an organo-silicon compound or a chromium complex.
2 A film former, which can typically be polyvinyl acetate latices, starch or emulsified polyester or epoxy resin. The film former protects the glass filaments during manufacture and subsequent handling, and gives integrity to the bundles of glass filaments.
3 A lubricant: certain silicones and acid amides can be used to lubricate filaments and aid their passage through and over guide points.
4 Anti-static agents: various organic and inorganic compounds are used to reduce the build-up of static.

For improvement of performance in composites the coupling agent is by far the most important component of the forming size.

At one end of the coupling agent molecules are reactive organic groups which are capable of typical organic reactions with organic polymers. Among the more common reactive organic groups are vinyl, alkylamine, methacryloxy, alkyl, mercapto and epoxy radicals. At the other end of the coupling agent are usually methoxy or ethoxy groups which attack hydroxy ($-OH$) groups on the glass surface to form links, as below:

Typical coupling agents are shown in Table 1.9 and react with the polymer via the active organic group as shown below for a vinyl terminated silane with polyester resin:

```
                              ⬡
                              |
                            —CH
                              |
                            CH₂\    /
                              HC
                            /   \        COOR—
                      —ROOC—CH       CH—CH—COOR'—
           CH₂                  \CH         |
Coupling    ‖      + Polyester    |
agent      CH                    CH₂
            |                     |
           Si—OEt    ⟶          Si—OEt
          /  \                  /  \
         O    O                O    O
Glass    |    |                |    |
         Si   Si               Si   Si
          \   /                 \   /
          O  O                  O  O
           \ /                   \ /
           Si                    Si
          / \                   / \
```

Not only do coupling agents improve the properties of glass fibre reinforced thermosetting resin systems, but they have also been shown to improve mechanical and electrical properties of many reinforced thermoplastic polymer systems.

1.4.1.3 Properties

Typical properties of glass fibre reinforcement are shown in Table 1.10 and compared with other fibres in Fig. 1.4, where it can be seen that glass fibres have higher elongation to failure than carbon and polyaramid fibres, but lower strengths and moduli.

The comparison between fibre stress–strain curves with a typical resin stress–strain curve (Fig. 1.4) is startling and in the composite performance section (1.5) it is shown how excellent fibre properties are put to use in considerably raising the performance of resins to form useful structural materials.

Figure 1.5 shows the relative specific tensile strengths of various fibres compared with steel, highlighting the very high strength to weight ratio of polyaramid and carbon fibres.

1.4.1.4 Commercially Available Forms of Glass Reinforcement

Glass reinforcement in strand form limits its use in fibre reinforced plastics to specific processes such as filament winding and pultrusion. It is the ease with which glass strands can be converted into various fabrics that has brought about the growth of its use in many other fabrication processes. The following forms of glass fibre are the most useful in the reinforced plastic industry.

Table 1.9 Typical silane linking (coupling) agents for various polymers

Chemical nomenclature	Chemical composition	Thermosetting polymer	Thermoplastic polymer
Vinyltrichlorosilane	$CH_2 = CH\text{-}SiCl_3$	Polyester	—
Vinyltriethoxysilane	$CH_2 = CH\text{-}Si\,(OC_2H_5)_3$	Polyester	—
Vinyl-tris (β methoxy-ethoxy)-silane	$CH_2 = CH\text{-}Si\,(OCH_2CH_2OCH_3)_3$	Polyester	—
γ-Methacryloxypropyl trimethoxysilane	$\overset{\displaystyle CH_3\ \ O}{\overset{\displaystyle \mid\ \ \parallel}{CH_2 = C\text{-}C\text{-}(CH_2)_3Si\text{-}(OCH_3)_3}}$	Polyester	Polystyrene, polyethylene ABS*, poly-propylene
β (3, 4-Epoxycyclo-hexyl ethyl-trimethoxy-silane	$O(C_6H_9)\ CH_2CH_2Si\text{-}(OCH_3)_3$	Polyester, epoxy	Polystyrene, ABS, SAN†
γ-Glycidoxypropyl-trimethoxy-silane	$\overset{\displaystyle /O\backslash}{CH_2\text{-}CH\text{-}O\text{-}(CH_2)_3Si(OCH_3)_3}$	Polyester, epoxy, melamine, phenolic	ABS, SAN
γ-Amino propyl triethoxy silane	$H_2N\text{-}(CH_2)_3\text{-}Si(OC_2H_5)_3$	Epoxy, melamine, phenolic	PVC, polycarbonate, polypropylene, polymethyl-methacrylate
N-bis (beta-hydroxy ethyl) gamma-amino-propyltri-ethoxy-silane	$(HOCH_2CH_2)_2N(CH_2)_3\ Si(OC_2H_5)_3$	Epoxy,	PVC, nylon, polysulphone
N-β (aminoethyl) gamma-aminopropyl trimethoxysilane	$H_2NCH_2CH_2NH(CH_2)_3\ Si(OCH_3)_3$	Epoxy, phenolic	—

* Acrylonitrile butadiene styrene
† Styrene-acrylonitrile

Continuous filament rovings consist of one or more strands of fibres parallel wound, without twist, into a spool or 'cheese'. The number of strands and spool size depend upon the end use.

In *chopped strand mat* (CSM) rovings are chopped and dispersed uniformly on a mat forming stage. The fibres are firmly bound in the mat but have excellent wet-out characteristics. The binders are usually of two forms – emulsion, which is based on polyvinyl acetate emulsion, and powder, which is based on bisphenol polyester powder.

CSM is supplied in a range of weights from 225 g/m^2 to 900 g/m^2 containing 3–6% binder by weight, and usually made to be polyester compatible. Hence, their use with epoxy resins is limited, but some types of mat are better than others in this respect, e.g. powder bonded mat.

Table 1.10 Typical properties of glass fibres

Property	Glass E	Glass S	Polyaramid	Carbon High strength	Carbon High modulus	Jute
Tensile strength (MPa)	1700	3100	3600	3300	2600	100(600*)
Tensile modulus (GPa)	70	86	124	237	345	12(57*)
Elongation (%)	3.5	4.0	2.5	1.34	0.74	—
Poisson's ratio	0.22	—	—	—	—	—
SG	2.56	2.49	1.44	1.81	1.86	1.20
Coefficient of thermal expansion (10^{-6}/°C)	5.0	5.6	− 2L + 59 T	− 1 L + 17 T		—
Coefficient of thermal conductivity (W/mK)	1.04	—	—	105 L		—
Specific heat (J/kg °C)	800	740	1420	710		—
Dielectric constant at 10^6 H$_Z$	6.33	5.34	—	—	—	—
Electrical resistivity (Ω m)	10^{13}	—	—	16×10^{-6}	10×10^{-6}	—
Power factor (loss tangent) at 10^6 H$_Z$	0.001	0.002	—	—	—	—
Refractive index	1.547	1.523	2.0 L 1.6 T	—	—	—

* From flexural tests. L, Longitudinal; T, Transverse.

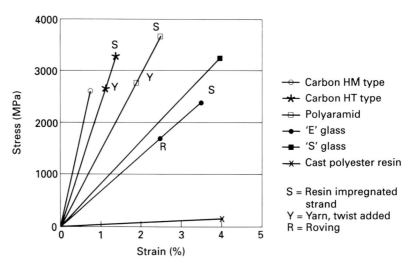

1.4 Stress–strain curves of reinforcing fibres.

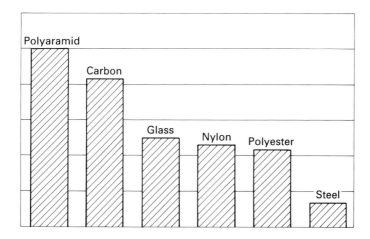

1.5 Specific tensile strengths of fibres compared with that of steel.

In addition to normal chopped strand mats, products are available which combine CSM with glass fibre tissue, woven rovings and unidirectional rovings to form *combination mats*.

Continuous Filament Mat (CFM) consists of multiple layers of continuous glass fibres deposited randomly in a swirl-like pattern. CFM is finding increasing use in matched-die moulding, where it can fill deep mould cavities, and in resin transfer moulding (RTM).

Woven glass fabrics can be divided into two classes: those prepared from rovings and those from yarn. Woven roving (WR) is made from 'E' glass rovings by weaving the untwisted rovings into fabric. Yarn-based cloths or fabrics are woven from continuous filament glass fibres which has been twisted before winding on to a bobbin. The use of size protects the yarn during weaving and the twist (20–40 turns/metre) enables uniform cloth to be woven.

Woven fabrics, whether made from yarn or rovings, are available in a number of different weaves, for example:

1   Plain weave, where each warp (along the roll) and weft (across the roll) thread passes over one thread and under the next, providing good distortion resistance and reproducible laminate thickness.
2   Twill weave, where the warp ends and weft picks which pass over each other can be varied, e.g. in a 2×1 twill, weft yarns pass over one and under two warp ends, while in a 2×2 twill, the weft yarns pass over two and under two warp ends, producing a regular diagonal pattern in the cloth. Twill weave cloth has good drapability.
3   Satin weave, where the number of warp ends and weft picks which pass over each other before interlacing is greater than with twill weave, and always with one crossing thread. Hence, one side of the fabric consists mostly of warp threads and the other weft threads.

Satin weave cloth has excellent drapability and imparts high properties to laminates.

4   Unidirectional weave, where warp threads are held together by fine weft threads such that the fabric is mostly unidirectional in structure. Maximum tensile properties are obtained in the warp direction with these fabrics.

Woven cloths and woven rovings are used in circuit boards, vehicle bodies, storage tanks, boats, etc., where improved mechanical perform- ance is required.

Woven rovings are supplied in weights from a few hundred g/m² to 1200 g/m².

1.4.2 Polyaramid Fibre Reinforcement

Polyaramid fibre is one of the most important man-made organic fibres ever developed and, because of its unique combination of properties, finds use in a wide variety of applications. The main feature of polyaramid fibre is high tensile strength with low density.

1.4.2.1 Manufacture

There is little detail available about the manufacturing process of polyaramid fibre, but the initial patent claims that poly-parabenzamide fibres can be produced by a solvent spinning process. Currently available polyaramid fibre is an aromatic polyamide called poly (paraphenylene terephthalamide):

$$\left[\ CO-\!\!\left\langle\,\bigcirc\,\right\rangle\!\!-CO-NH-\!\!\left\langle\,\bigcirc\,\right\rangle\!\!-NH\ \right]$$

The fibres are produced by extrusion and spinning processes by holding the polymer in a suitable solvent at between $-50\ ^\circ$C and $-80\ ^\circ$C before extrusion into a hot walled cylinder at 200 °C where the solvent evaporates. The fibre is then stretched and drawn, resulting in the alignment of the stiff polymer molecules with the fibre axis.

The molecules form rigid planar sheets with chain-extended mole- cules hydrogen bonded together. The sheets are stacked to form a crys- talline array, but there is only weak bonding between the sheets; the weaknesses in the structure probably account for the very low resistance to compressive forces.

1.4.2.2 Properties

The tensile properties of polyaramid fibre are compared with other fibres in Fig. 1.4. More detailed properties are given in Table 1.10.

Polyaramid fibres are over five times stronger than steel on a weight for weight basis, with excellent thermal and dimensional stability, excellent wear resistance and good heat resistance. Carbon and glass fibres are almost completely brittle and fracture without reduction in cross-sectional area. However, polyaramid fibres fracture in a ductile manner, and fracture involves fibrillation of the fibres.

Polyaramid fibres, unlike carbon and glass fibres, are more susceptible to fibre breaking in bending by a compressive mode, because of their low compressive strength. Bending the fibres produces high surface compressive stresses as well as tensile stresses, but long before the bending curvature is sufficient to cause tensile fracture, the compressive region of the fibre undergoes yielding by development of deformation bands. However, the many advantages of this type of fibre can, with adequate thought, be used whilst guarding against the disadvantage of low compressive strength. In fact, combination with other fibres in the form of hybrid fabrics often provides an acceptable compromise to take advantage of the unique properties of polyaramid fibres.

1.4.3 Carbon Fibre Reinforcement

Carbon fibre is the most expensive of the more common reinforcements, but in aerospace applications the combination of excellent performance characteristics coupled with light weight make it an indispensable reinforcement with cost being of secondary importance.

1.4.3.1 Manufacture

There are a number of precursors for the development of carbon fibres, including cellulose fibre, polyacrylonitrile fibre, lignin and pitch. Today, high strength, high modulus carbon fibres are prepared in large quantities from both cellulose and polyacrylonitrile precursors. Lower strength fibres can be produced from a special hydrocarbon pitch. In all processes, carbon fibres are produced by controlled oxidation and carbonisation of the precursor fibre at temperatures up to 2600 °C, resulting in high strength fibre. Increasing the temperature to 3000 °C results in the conversion of high strength fibre to high modulus graphite fibre. The conversion of polyacrylontrile fibre to carbon fibre is more efficient than that using cellulose fibre, owing to the higher carbon content of the precursor.

1.4.3.2 Properties

The tensile properties of carbon fibres are compared with other fibres in Fig. 1.4 and it is very clear that carbon fibre reinforced resins offer much more to the designer than many other materials in current use, especially in terms of strength and rigidity. The properties of carbon fibres from various precursors are given in Table 1.11.

Carbon fibres are generally surface-treated to improve bonding

Table 1.11 Typical properties of carbon fibres from various precursors

Fibre precursor	Fibre type	Diameter (µm)	SG	Tensile strength (MPa)	Tensile modulus (GPa)	Elongation (%)
Polyacrylonitrile	Carbon	8	1.76	3300	230	1.34
Polyacrylonitrile	Graphite	8	1.87	2600	345	0.74
Cellulose	Graphite	6.6	1.67	2000	390	0.5
Lignin	Carbon	10–15	1.5	600	—	1.5
Hydrocarbon pitch	Carbon	10.5	1.6	1030	—	2.5

Coefficient of thermal expansion (10^{-6}/C):	−1 longitudinally + 17 transversely
Coefficient of thermal conductivity (W/mK):	105 longitudinally
Electrical resistivity (Ω m):	$10\text{--}16 \times 10^{-6}$
Specific heat (J/kg °C):	710

between fibre and matrix. Sizing with a resin also improves handling and prevents damage during processing.

Carbon fibres are supplied in a number of different forms, from continuous filament tows to chopped fibre mat. The highest strength and modulus are obtained by using unidirectional reinforcement. Twist-free tows of continuous filament carbon contain 5000–10 000 individual filaments, which can be woven into woven roving and hybrid fabrics with glass fibre and polyaramid fibre.

1.4.4 Other Reinforcing Fibres

There is little use of fibres other than glass, polyaramid or carbon for the manufacture of fibre reinforced plastics. However, other fibres are available and have been investigated for other reasons. Some examples are as follows:

1 Polyester fibres are used in the manufacture of surfacing tissues and if used for structural reinforcement produce laminates with very high impact resistance, excellent chemical resistance and excellent abrasion resistance.
2 Jute fibres are cheap, readily available, naturally occurring fibres which are used in woven cloth and yarn form in some Third World countries.
3 Sisal fibres are inexpensive, naturally occurring fibres used in phenolic based DMCs, but rarely with polyester or epoxy resins.
4 Nylon fabrics are used to reinforce epoxy resins when high impact, abrasion resistant and chemical resistant laminates are required. They are mostly used in combination with glass reinforcements.

5 Boron fibres are used to reinforce epoxy resins for specialised aerospace applications, but their very high cost limits their use.

Reinforcing fibres are carefully chosen for each application to impart the desired properties in the component by placing fibres in the directions of maximum stress. In some instances, requirements for light weight call for the use of fibres other than glass (such as those based on carbon and polyaramid) but cost will also become a vital factor, with polyaramid fibres some 15 times the cost of glass fibres and carbon fibres 40 times the cost of glass fibres.

Whatever the fibre used in a given construction, the resultant performance of the composite will be both unique and unlikely to be achieved with any other combination of fibre and resin for the same weight and cost.

The next section discusses the performance of fibre reinforced composites.

1.5 Properties of Composite Materials

The mechanical behaviour of composite materials is not governed by the fibres alone, but by a synergy between the fibres and the matrix. However, the stiffness of the composite is very much a function of the reinforcing material although tensile strength is a product of the fibre–matrix synergy, in that when fibres without a surrounding matrix are stressed, the failure of a single fibre eliminates it as a load carrier; this stress has, then, to be carried by the remaining fibres, moving them closer to failure, but fractured fibres embedded in a matrix still retain a mechanical function by supporting load, at least until their length is below the critical fibre length for load transfer from the matrix.

The proportion of reinforcement present in a composite also has a major effect on properties, as does fibre type and fibre orientation. Fibre content is usually expressed in terms of a weight fraction, a volume fraction or a resin : fibre ratio. Typical resin : fibre ratio data are shown in Fig. 1.6 and 1.7 for both unfilled and filled resin.

$$\text{Fibre weight fraction } W_f = \frac{1}{1 + R}$$

where R = resin : fibre ratio.

$$\text{Fibre volume fraction } V_f = \frac{1}{1 + R\,(d_f/d_m)}$$

where d_f = density of fibre and d_m = density of curved matrix.

The individual fibre curves, in Fig. 1.6 and 1.7, are designed to be read in conjunction with the volume fraction scale. Conversion from weight fraction or percentage by weight to volume fraction is achieved

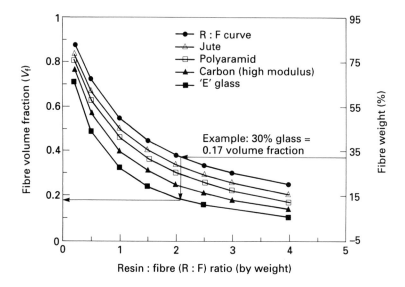

1.6 Resin–fibre relationship with unfilled polyester resin of SG = 1.23.

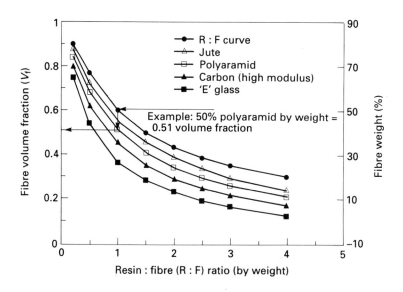

1.7 Resin–fibre relationship with filled polyester resin of SG = 1.50.

via the intermediate resin/fibre (R : F) ratio as shown in Fig. 1.6, for 50% polyaramid by weight. The top curve shows that the R : F ratio is 1.0 and projecting from the polyaramid curve at R : F=1, to the volume fraction scale, the volume fraction is found to be 0.46. Volume fraction can also be calculated from:

$$V_f = \frac{d - d_m}{d_f - d_m}$$

where d = laminate density, d_m = matrix density and d_f = fibre density which is derived from the law of mixtures:

$$d = V_f d_f + V_m d_m$$

where $V_m = 1 - V_f$.

Figure 1.8 shows the relationship between laminate specific gravity (SG) and fibre fraction for various fibres with filled and unfilled polyester resin.

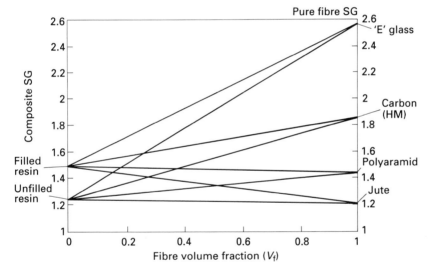

1.8 Relationship between composite SG and fibre fraction for various fibres with unfilled and filled polyester resin.

1.5.1 Measured Properties

The data given in Table 1.12 are based on measured properties for various types of laminates covering mechanical performance in tension, flexure, compression and shear. Comparisons are given with steel and aluminium alloy. Other general properties of laminates are given in Table 1.13.

It must be recognised that a considerable spread in properties can

Table 1.12 Typical measured mechanical properties of composite materials compared with steel and aluminium alloy

Reinforcement	CSM (PB)	WR	Woven cloth (satin weave)	Filament wound — Hoop wound	Filament wound — Angle-ply 55°	Combination mat — UD-CSM	Combination mat — WR-CSM	SMC
					Glass fibre			
Resin	Ortho polyester	Iso polyester	Iso polyester	Iso polyester	Iso polyester	Iso polyester	Iso polyester	Iso polyester
Post-cure	16 h 40 °C	16 h 40 °C	28 day room temperature	3 h 80 °C	3 h 80 °C	16 h 60 °C	16 h 40 °C	Hot-cure
Fibre volume fraction	0.17	0.32	0.37	0.44	0.47	—	—	24% glass by wt
R:F ratio (by weight)	2.33 : 1	1 : 1	0.81 : 1	—	—	1.3 : 1	1.16 : 1	—
Direction of stressing	—	0°/90°	0°/90°	0° / 90°	Hoop / Axial	0° / 90°	0°/90°	—
Property								
SG	1.46	1.7	1.7	1.83 / —	— / —	— / —	—	1.87
Tensile strength (MPa)	110	220	260	650 / 25	340 / 450	300 / 25	180	70
Tensile modulus (GPa)	8	14	17	30 / 10	20 / 29	16 / 7	12.2	10
Tensile elongation to failure (%)	1.6	1.7	—	— / —	— / —	2.0 / 0.8	1.8	1.4
Tensile Poisson ratio	0.32	0.14	0.17	0.26 / 0.11	0.62 / —	— / —	—	—

Property											
Compression strength (MPa)	150	230	210	800	120	—	—	250	110	180	120
Compression modulus (GPa)	8	15	19	36	10	—	—	—	—	—	—
Compression strain to break (%)	—	—	—	2.6	1.2	—	—	—	—	—	—
Compression Poisson ratio	0.42	0.25	0.23	0.30	0.11	—	—	—	—	—	—
Flexural strength (MPa)	190	270	480	800	—	—	—	—	—	320	140
Flexural modulus (GPa)	7	14	17	35	—	—	—	—	—	11	8
Shear strength (MPa)	80	90	70	50	—	—	—	—	—	—	70
Shear modulus (GPa)	3	3.3	—	4	—	—	—	—	—	—	3.8
Short beam shear strength (MPa)	20	25	36	50	—	—	—	—	—	—	19
Interlaminar lap shear strength (MPa)	5	7	—	—	—	—	—	—	—	7	—

Table 1.12 cont.

Reinforcement	Polyaramid		Carbon	Mild steel	Aluminium alloy
	WR	UD roving	UD High modulus		
Resin	Iso polyester	Epoxy	Epoxy	—	—
Post-cure	28 day room temp.	180 °C	180 °C	—	—
Fibre volume fraction	—	0.60	0.60	—	—
R:F ratio (by weight)	1:1	—	—	—	—
Direction of stressing	0° 45°	0° 90°	0° 90°	—	—
Property					
SG	—	1.38	1.60	7.8	2.8
Tensile strength (MPa)	390 66	1380 30	1260 40	450	300
Tensile modulus (GPa)	24 5.5	76 5.5	200 7.6	207	70
Tensile elongation to failure (%)	1.9	1.8 0.6	0.5 0.5	45	—
Tensile Poisson ratio	—	0.34	0.30 0.30	0.27	0.32

Compression strength (MPa)	86	63	276	138	840	140	—	—
Compression modulus (GPa)	20	6	76	5.5	190	7	—	—
Compression strain to break (%)	—	—	—	—	0.40	2.5	—	—
Compression Poisson ratio	—	—	—	—	—	—	—	—
Flexural strength (MPa)	190	130	620	—	1070	45	—	—
Flexural modulus (GPa)	16	4.3	76	—	190	—	—	—
Shear strength (MPa)	—	—	60	60	65	65	330	180
Shear modulus (GPa)	—	—	2.1	2.1	5.5	5.5	80	26
Short beam shear strength (MPa)	17	15	41	83	55	87	—	—
Interlaminar lap shear strength (MPa)	10	8	—	—	—	—	—	—

Table 1.13 Other general properties of composites compared with steel and aluminium alloy

	Material									
	Random rein-forced GFRP	SMC 24% glass	UD GFRP	WR GFRP	UD polyaramid -epoxy		UD carbon -epoxy		Mild steel	Alum-inium alloy
					0°	90°	0°	90°		
Charpy impact strength (kJ/m²) (unnotched)	75	60	250	—	—	—	—	—	50	25
Coefficient of thermal expansion (10^{-6}/°C)	30	—	10*	15	−4	57	−0.5	25	12	23
Thermal conductivity (W/m °C)	0.2	—	0.3	0.24	1.7	0.14	34	0.8	50	200
Maximum working temperature (°C) (depending upon type of resin and loading conditions)	175	—	250	—	—	—	—	—	400	200
Specific tensile strength (MPa)	75	38	417	150	1000	20	800	25	31	154
Specific tensile modulus (GPa)	5.5	5.0	25	10	55	4	125	5	27	26

* Longitudinally.

occur with most types of reinforcement owing to factors such as resin uptake, lay-up conditions, post-curing, etc. In fact, just changing resin type or CSM type, at the same resin to fibre ratio, can result in significant changes in properties, as shown in Fig. 1.9.

It is well recognised that post-cure improves resin properties and some composite properties. However, since many structures are cured at room temperature, it is important to compare the performance of different resins under room temperature cured conditions. As can be seen in Fig. 1.10, 1.11 and 1.12, there is no benefit in terms of mechanical performance in using a cold-cure epoxy or a vinyl ester resin instead of an isophthalic polyester resin under cold-cure conditions. In fact, since the optimum benefits of the more 'exotic' expensive resins are only achieved after high temperature post-cure, there appears to be little point in paying the premium for such resins, unless it is intended to use a high temperature post-cure cycle during the production cycle.

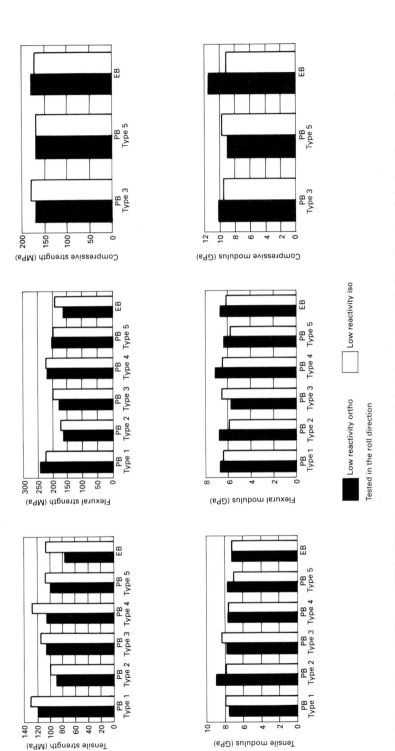

1.9 Comparison of different CSM (30% fibre content by weight) with an orthophthalic and an isophthalic polyester resin (PB = powder bonded, EB = emulsion bonded).

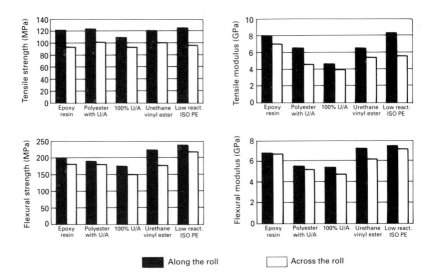

1.10 Comparison of tensile and flexural properties of CSM with various resin laminates with the same glass content (U/A = urethane/acrylate).

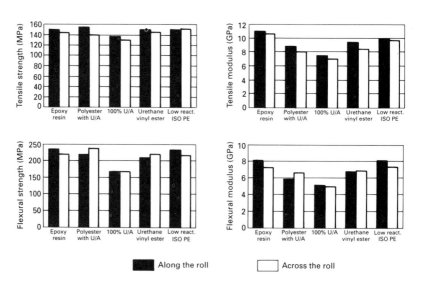

1.11 Comparison of tensile and flexural properties of WR–CSM with various resin laminates and the same glass content.

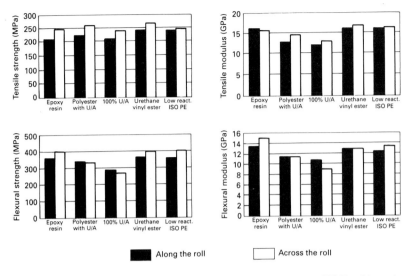

1.12 Comparison of tensile and flexural properties of WR with various resin laminates with the same glass content.

1.5.2 Variation of Composite Properties with Temperature

In some applications, the retention of property at elevated temperature is an important parameter. Tables 1.14 and 1.15 illustrate the percentage retention of room temperature properties for a fully cured epoxy–glass cloth reinforced laminate and a fully cured isophthalic polyester–chopped strand mat laminate.

At low temperatures there is often property improvement compared with room temperature, but as temperature increases and approaches the heat deflection temperature of the resin, so properties are dramatically reduced.

1.5.3 Prediction of Properties

The basic measured laminate properties are a reasonable first guide to a laminate construction for a particular application. However, an alternative approach is to calculate or predict the properties of chosen laminate constructions.

A theoretical 'micromechanics' approach has been extensively developed for unidirectional laminae in the high performance aerospace field for the prediction of elastic modulus, strength and other properties.

Table 1.14 Percentage retention of tensile properties at various temperatures for a glass cloth reinforced epoxy resin

Property	Temperature (°C)			
	−54	24	71	149
Tensile strength	118	100	78	74
Tensile modulus	106	100	86	76
Elongation	123	100	82	74

Table 1.15 Percentage retention of tensile properties at various temperatures for a CSM reinforced isophthalic polyester resin with an HDT of 116 °C

Property	Temperature (°C)					
	−68	−40	23	70	130	225
Tensile strength	98	98	100	104	56	33
Tensile modulus	105	94	100	65	34	—
Elongation	100	129	100	143	143	—

1.5.3.1 Tensile Properties

The calculations are more successful for modulus than for strength predictions – the simple law of mixture is reasonably accurate for tensile modulus using the equation:

$$E = \beta\, E_f V_f + E_m V_m$$

where E_f and V_f are the modulus and volume fraction of the fibre respectively; E_m and V_m are the modulus and volume fraction of the matrix respectively; β = the fibre efficiency factor (= 1.0 for unidirectional reinforcement (U/D); 0.5 for balanced bidirectional reinforcement (WR); 0.375 for chopped strand mat reinforcement (CSM)).

Figure 1.13 shows modulus values, calculated using the law of mixtures, for U/D and WR laminates, made from both glass and polyaramid fibre, as well as for glass CSM reinforcement. Comparison with measured typical values, given in Table 1.12 shows tolerably good agreement.

A similar equation, including the same efficiency factor, can be used for strength prediction, but one major difficulty is the choice of the effective ultimate fibre strength to be used in the calculations, because in the laminate, some fibres lose strength rather rapidly as a result of handling and processing. The law of mixtures, when applied to ultimate strength calculation, also implies equal strain at failure in both fibre and resin, which will only be approximately true with the more

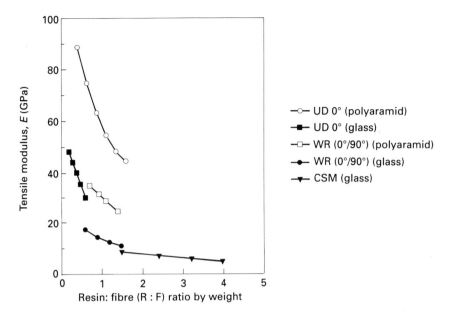

1.13 Predicted effect of fibre type and content on tensile modulus using law of mixtures.

common reinforcements and matrices in use for the manufacture of structural composites.

In view of these uncertainties, an empirical approach to strength prediction is probably a preferred option. Such a method is based on strength and stiffness factors obtained from measurement of laminate performance for different materials made up of various fibres and fibre contents. Useful factors are given in Table 1.16.

It should be noted that values for both unit strength and extensibility (a tensile stiffness/modulus related property) are expressed as force per unit width of laminate for unit weight/m² of reinforcement. The implication is that laminate tensile properties are directly proportional to reinforcement weight and orientation, but independent of resin weight and laminate thickness.

An advantage of this approach is that strength and stiffnesses of mixed laminates are easily calculated by addition of the values of individual plies (layers). Strength values are not strictly additive unless all plies fail at the same strain, but the procedure is probably sufficiently accurate for many design purposes, especially when subsequently backed up by measurements on trial laminates; a desirable if not mandatory requirement.

Figures 1.14 and 1.15 show the tensile strengths and moduli of laminates with varying resin : fibre ratio calculated using the factors given

Table 1.16 Minimum properties of reinforced laminate plies (layers)

Reinforcement	Ultimate unit tensile strength (N/mm width per kg/m² of reinforcement)	Extensibility (unit modulus – N/mm width per kg/m² of reinforcement)
Glass		
CSM	200	14 000
WR (0°, 90°) (balanced)	250	16 000
UD roving	430	25 000
UD (continuous filament – 0°)	500	28 000
UD – 90°	0	8 400
Polyaramid		
Woven (0°, 90°) (balanced)	600	35 000
UD (0°)	1200	70 000

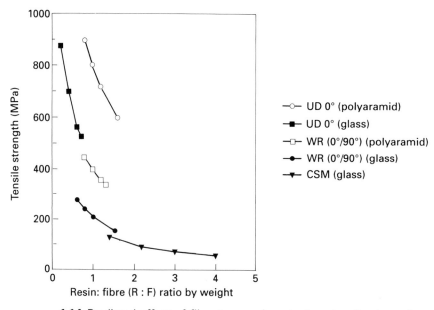

1.14 Predicted effect of fibre type and content on tensile strength using property data given by Table 1.16.

in Table 1.16. In unidirectional laminates the transverse modulus is not negligible, unlike transverse strength, and should always be considered at a level of about 30% of the longitudinal value.

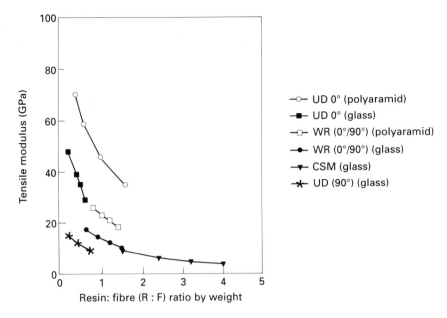

1.15 Predicted effect of fibre type and content on tensile modulus using property data given by Table 1.16.

The data in Fig. 1.14 and 1.15 are given in conventional material property units (i.e. per unit cross-sectional area) by including theoretical laminate thickness in the calculations:

$$\text{Thickness } t \text{ (in mm)} = \frac{W_f}{d_f} + \frac{W_m}{d_m}$$

where W_f and d_f are the weight and density of the fibre in kg/m² and g/cm³ respectively; W_m and d_m are the weight and density of the matrix in kg/m² and g/cm³ respectively.

The calculations of strength and stiffness values for mixed laminates are aided by the use of Fig. 1.16, 1.17 and 1.18, where Fig. 1.16 and 1.17 show the minimum tensile strength and stiffness (per width) expected for varying weights of reinforcement up to 10 kg/m² and Fig. 1.18 shows the predicted thickness for varying weights of reinforcement and resin up to 10 kg/m².

Two worked examples of property prediction, using the approach discussed here, are now given.

Example 1:
Consider a laminate containing five layers of polyaramid woven rovings (450 g/m² per layer) and five layers of CSM (225g/m² per layer) laid up using unfilled resin at an overall resin : fibre ratio of 1.34 : 1

weight of woven polyaramid	= 5 × 0.45
	= 2.25 kg/m²
from Fig. 1.16 strength	= 1350 N/mm
from Fig. 1.17 stiffness	= 80 kN/mm
weight of CSM glass	= 5 × 0.225
	= 1.125 kg/m²
from Fig. 1.16 strength	= 225 N/mm
from Fig. 1.17 stiffness	= 15.5 kN/mm
Total strength (per unit width)	= 1350 + 225
	= 1575 N/mm
Total stiffness (per unit width)	= 80 + 15.5
	= 95.5 kN/mm

The measured properties of a laminate with this specification were:

tensile strength (per unit width) = 1823 N/mm
tensile stiffness (per unit width) = 109 kN/mm

Since the estimated properties were the minimum expected, the agreement is considered reasonable.

weight of resin required = (2.25 + 1.125) × 1.34 kg/m²
= 4.522 kg/m²

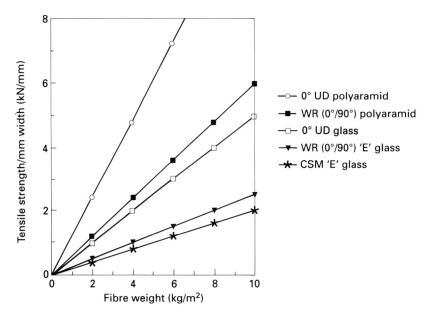

1.16 Predicted minimum laminate tensile strength as a function of fibre type and content.

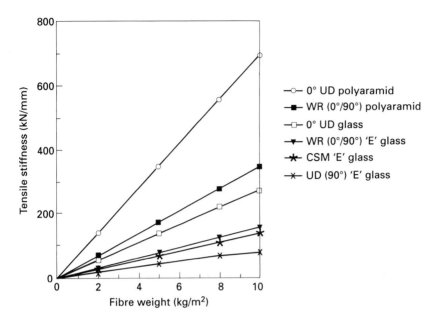

1.17 Predicted minimum laminate tensile stiffness as a function of fibre type and content.

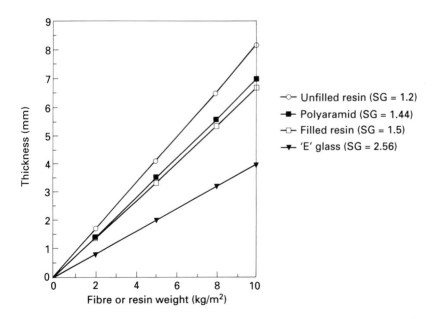

1.18 Predicted thickness for various weights of fibres and cast resin.

The thickness of the laminate can now be calculated as follows, using Fig. 1.18:

thickness due to polyaramid	= 1.58 mm
thickness due to glass	= 0.45 mm
thickness due to resin	= 3.70 mm
total thickness	= 5.73 mm
predicted tensile strength	= 1575/5.73 = 275 MPa
predicted tensile modulus	= 95.5/5.73 = 16.7 GPa

These values all compared favourably with the measured values of 6.0 mm, 296 MPa and 17.7 GPa respectively.

Example 2:
A rather more complex case, involving different properties in the two principal directions is shown below. The laminate is made up as follows:

three layers of 600 : 150 g/m² UD/CSM fabric at 1 : 1 R : F ratio;
one layer of 600 : 300 g/m² WR/CSM fabric at 1.3 : 1 R : F ratio;
one layer of 300 : 300 g/m² WR/CSM fabric at 1.4 : 1 R : F ratio;
one layer of 300 g/m² CSM at 2 : 1 R : F ratio.

The UD/CSM fabric was laid up such that in two layers the unidirectional fibres were in the main (0°) direction and in the transverse (90°) direction for the third layer.

Table 1.17 gives details of each layer in the laminate under consideration, together with predicted data. These data are compared with the measured data in Table 1.18. The predicted and measured data give very good agreement. Some variation can be accounted for by the higher than specified resin content in the test laminate, which is reflected in laminate thickness. Hence, to avoid the misleading aspects of this variable, it is more useful to work with unit properties (composite properties) rather than material properties.

1.5.3.2 Variation of Properties with Angle
The prediction methods already outlined have been concerned only with the main 0° and 90° directions of a laminate, i.e. the warp (along the roll) and weft (across the roll) directions of woven roving and woven yarn reinforcements, or the fibre and transverse directions of unidirectional reinforcement. At angles between the 0° and 90° directions, the properties can vary substantially, except for chopped strand mat (isotropic in the plane) laminates, which have equal properties in all directions.

For the orthotropic laminates, which include unidirectional, woven roving and plain weave cloth laminates, elastic properties (tensile modulus, Poisson ratio and shear modulus) can be calculated for any angle once these properties are known for the 0° and 90° directions; the notation used here is the one adopted in Section 5.2 and shown in

Table 1.17 Details of each layer and predicted properties for a specific laminate (using Fig. 1.16–1.18)

Layer No.	Layer type	Glass wt (g/m²)	Layer composition	R : F ratio	Resin wt (kg/m²)	Unit tensile strength (N/mm) 0°	90°	Tensile stiffness (kN/mm) 0°	90°
1	Mat	300	CSM	2 : 1	0.60	60	60	4.20	4.20
2	WR–CSM	300	CSM-WR	1.4 : 1	0.84	60	60	4.20	4.20
	combination	300	—	—	—	75	75	4.86	4.86
3	UD–CSM	150	CSM-UD	1 : 1	0.75	30	30	2.10	2.10
	combination	600	0°	—	—	300	0	16.80	5.04
4	UD–CSM	150	CSM-UD	1 : 1	0.75	30	30	2.10	2.10
	combination	600	0°	—	—	300	0	16.80	5.04
5	UD–CSM	150	CSM-UD	1 : 1	0.75	30	30	2.10	2.10
	combination	600	90°	—	—	0	300	5.04	16.80
6	WR–CSM	300	CSM-WR	1.3 : 1	1.17	60	60	4.20	4.20
	combination	600	—	—	—	150	150	9.72	9.72
Totals		4050			4.86	1095	795	72.12	60.36

Predicted value of materials property based on a total predicted thickness of 5.53 mm (1.58 mm for reinforcement, 3.95 mm for resin).

	0°	90°	0°	90°
	198 MPa	144 MPa	13.0 GPa	10.9 GPa

Table 1.18 Comparison between predicted and measured laminate criteria for the example under consideration

Property	Predicted 0°	90°	Measured 0°	90°
Unite tensile strength (N/mm)	1095	795	1350	874
Tensile stiffness (kN/mm)	72.1	60.4	72.2	65.5
Tensile strength (MPa)	198	144	231	144
Tensile modulus (GPa)	13.0	10.9	12.3	10.8
Glass weight (kg/m²)	4.05		3.74	
Resin weight (kg/m²)	4.86		5.38	
Laminate thickness (mm)	5.53		6.00	

Fig. 5.3. The modulus of elasticity, E, due to the stress applied in a direction at an angle of theta to the warp direction is obtained from:

$$\frac{E_{11}}{E_\theta} = \cos 4\theta + \frac{E_{11}}{E_{22}} \sin 4\theta + \frac{1}{4} \left[\frac{E_{11}}{G_{12}} - 2 v_{12} \right] \sin^2 2\theta$$

The Poisson ratio is obtained from:

$$v_{12} = \frac{E_{11}}{E_{22}} \left[v_{12} - \frac{1}{4} (1 + 2 v_{12} + \frac{E_{11}}{E_{22}} - \frac{E_{11}}{G_{12}}) \sin^2 2\theta \right]$$

The ultimate tensile strength is obtained from:

$$\frac{1}{(\sigma_\theta)^2} = \frac{\cos 4\theta}{(\sigma_{11}{}^*)^2} + \frac{\sin^4\theta}{(\sigma_{22}{}^*)^2} + \frac{\sin^2\theta \cos^2\theta}{(\sigma_{12}{}^*)^2}$$

where:

E_{11} = the tensile modulus in the $0°$ direction;
E_{22} = the tensile modulus at $90°$;
G_{12} = the shear modulus;
v_{12} = Poisson ratio (strain in the transverse direction due to load in the $0°$ direction);
$\sigma_{11}{}^*$ = the ultimate tensile strength in the $0°$ direction;
$\sigma_{22}{}^*$ = the transverse tensile strength;
$\sigma_{12}{}^*$ = the shear strength;
E_θ = the tensile modulus at the angle θ;
σ_θ = the tensile strength at the angle θ.

Figure 1.19 shows the predicted curve for the variation in tensile modulus between $0°$ and $90°$ for a unidirectional laminate and the good agreement with the experimental points at different angles. For $0°$–$90°$ woven roving and woven cloth laminates, minimum values for tensile and compressive moduli are predicted at $45°$ as expected, with values in the range 40–60% of the $0°$ and $90°$ values. Shear modulus and Poisson ratio have maximum values at $45°$ in WR laminates, at levels two to three times as high as at $0°$ and $90°$.

Typical measured data are shown in Fig. 1.20 for balanced bidirectional and tridirectional stitched fabrics of glass, glass–polyaramid hybrid and glass–carbon hybrid.

Figure 1.21 shows typical curves for bidirectional woven roving and WR/CSM combination mat. The properties for the 1500 g/m² WR and 1500/300 g/m² combination mat are higher in the $90°$ direction because of the unbalanced nature of the fabric.

Also shown in Fig. 1.21 are typical theoretical curves for strength and modulus for a WR/CSM combination product at a fibre volume fraction of 0.33. While the curve for modulus clearly obeys the law of mixture, the strength curve goes through a minimum, indicating a need for a minimum WR content before any benefit of its presence is

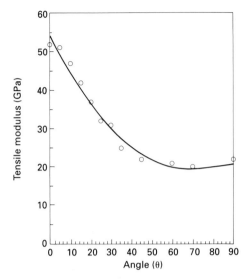

1.19 Experimental values of the tensile modulus compared with the theoretical curve for a unidirectional laminate, from 0° to 90°.

felt. Similar strength predictions can be made for hybrid fabrics of glass/polyaramid, glass–carbon and carbon–polyaramid. There is a critical volume fibre content (of the stiffer fibre) before a strength benefit in a laminate occurs.

1.5.3.3 Flexural Properties
In tension the strength and stiffness of laminates are provided almost entirely by the reinforcement, with the resin playing a secondary role. In compression, however, although the reinforcement dominates stiffness, it has much less influence on strength where the resin compressive strength plays a critical part. Similarly, in flexure the resin contribution to properties cannot be ignored, influencing both the bending moment at break and flexural rigidity powerfully through its effect on laminate thickness.

For glass fibre reinforced laminates the bend moment at break (per unit width of laminate) is given by $\sigma^* t^2/6$, where σ^* = ultimate tensile strength and t = thickness of laminate.

Since σ^* can be calculated approximately from the law of mixtures, the following holds true:

$$\sigma^* = \beta\sigma^*_f V_f + \sigma^*_m V_m$$

where σ^*_f and V_f are the ultimate tensile strength of the fibre and volume fraction of fibre respectively; σ^*_m and V_m are the corresponding matrix parameters; and β is the efficiency factor.

Bidirectional fabric   Tridirectional fabric

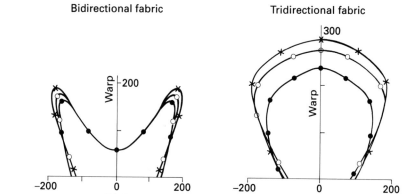

(a)   Tensile strength (MPa)

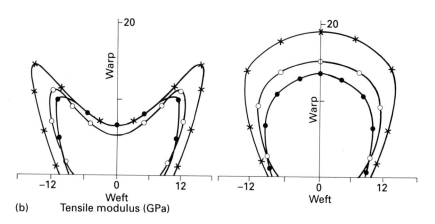

(b)   Tensile modulus (GPa)

- ● 400 g/m² glass in principal directions
- ○ 400 g/m² glass–polyaramid in principal directions
- x 400 g/m² glass–carbon in principal directions

1.20 Measured tensile properties of multidirectional hybrid fabrics.

Thickness t is calculated from the weights of fibre and resin and their densities, thus:

$$t = \frac{W_f}{d_f} + \frac{W_m}{d_m}$$

as before, where W is in kg/m² and d is in g/cm³.

Figure 1.22 shows how the bending moment at failure increases with weight of glass reinforcement and resin : fibre (R : F) ratio for glass

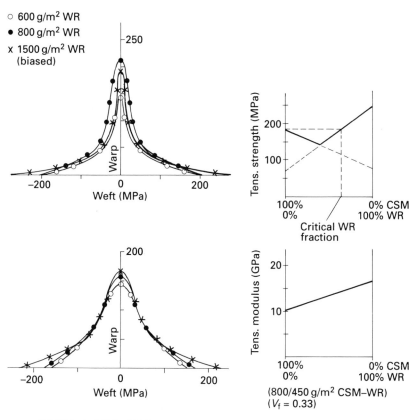

○ 600 g/m² WR
● 800 g/m² WR
x 1500 g/m² WR
(biased)

○ 600/300 g/m² WR–CSM
● 800/300 g/m² WR–CSM
x 1500/300 g/m² WR (biased)–CSM

1.21 Tensile strength of CSM–WR and WR laminates in different directions and theoretical curves in 0°/90° directions for CSM–WR fabric.

reinforced laminates. For a given weight of glass the bend moment at break is highest for the CSM laminates at all but the lowest resin : fibre ratios; this, of course, is due to the t^2 term in the expression for bending moment at break and the effect of the resin in adding to laminate thickness.

The calculation used here for assessing the ultimate bending moment of glass reinforced laminates is not applicable to polyaramid reinforced laminates, because their much lower strength in compression compared with their strength in tension results in a different mode of failure – usually at the compressive face rather than the tensile face.

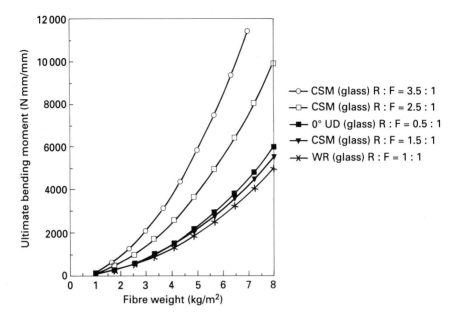

1.22 The effect of glass content, glass type and resin (unfilled) content on bend moment at break.

The flexural rigidity (per unit width of laminate) is given by $Et^3/12$, where E = tensile modulus and t = laminate thickness. Here, E can be calculated by the law of mixtures, and t from the weights and densities of the fibre and the matrix.

Figure 1.23 shows how flexural rigidity increases with weight of reinforcement for seven types of laminate. The same calculation can be applied equally to polyaramid fibre and glass fibre, since only small elastic deformations, not rupture properties, are involved.

The effect of increased resin content is even more marked than for bend moment at break, because a t^3 term is involved. Although bending stiffness is proportional to the modulus of elasticity, even laminates made of high modulus directional glass reinforcement cannot usually compensate for the effect of t^3 in lower modulus CSM materials to achieve a given bending stiffness.

1.5.3.4 Properties of Sandwich Construction

The full theory of sandwich construction is somewhat complex and a complete subject in its own right. However, its brief inclusion here is essential, as sandwich constructions are attracting more and more interest in the production of lighter and lighter structural composites. In Section 5.5 a further discussion of the sandwich construction systems is given.

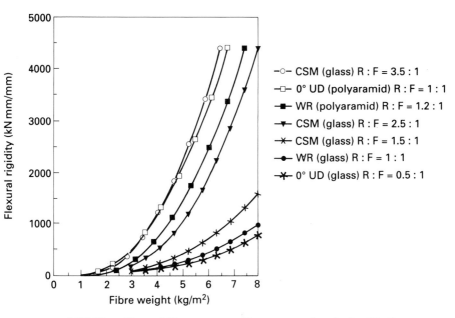

1.23 The effect of fibre content, fibre type and resin (unfilled) content on flexural rigidity.

For the sake of this exercise, the limitation will be for thin faces (less than 20% of the core thickness) and weak core materials (such as low density foamed plastics). Calculations of strengths and stiffnesses of such systems can be carried out using relatively simple expressions involving the parameters shown in Fig. 1.24.

The Bending Strength (face failure) of the Sandwich Beam (see also Section 5.5.1)
The value of the bending moment of a sandwich beam is best

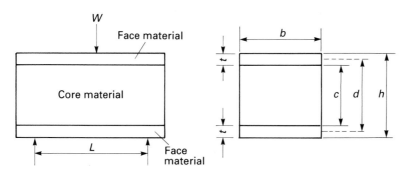

1.24 Parameters for sandwich composite calculations.

expressed as the bending moment at failure. When failure occurs in one face in tension, as is normal in pure bending under four-point loading, or long span three-point loading, the bend moment at break can be calculated from the ultimate tensile stress of the skins. Thus, the stress σ produced in the skins by an applied moment M is:

$$\sigma_b = \frac{M}{bdt}$$

The ultimate bending moment of the sandwich construction is therefore $M^* = bdt\sigma^*$ where σ^* is the ultimate tensile stress in the face material and $t\sigma^*$ is the ultimate tensile unit strength of the face material. (Minimum values for various materials are given in Table 1.16).

Using this information, Fig. 1.25 and 1.26 have been drawn for sandwich constructions made with glass fibre reinforced plastic (GFRP) skins from CSM and WR respectively. They show bend moment at break per mm width, plotted against the reinforcement weight in each skin for four different core thicknesses, c.

The influence of the two main variables in the sandwich construction (skin reinforcement weight and core thickness) is clearly shown in Fig. 1.25 and 1.26. For this exercise a resin : fibre (R : F) ratio of 2.33 : 1 was assumed for the CSM skins and 1 : 1 for the WR face. The R : F ratio of the faces has only a marginal effect on the bending

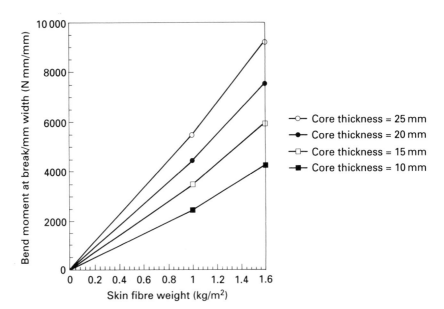

1.25 The effect of CSM skin construction (at R : F = 2.33 : 1) and core thickness on bend moment at break for balanced double skinned sandwich laminates.

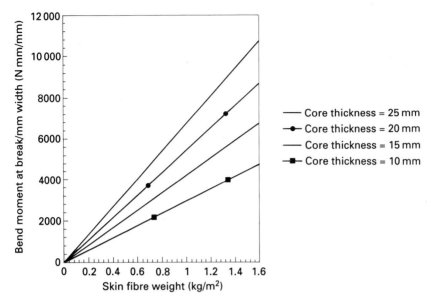

1.26 The effect of WR (glass) skin construction (at R : F = 1 : 1) and core thickness on bend moment at break for balanced double skinned sandwich laminates.

moment at failure of a sandwich construction, because resin variation in the faces has only a minimum effect on the overall thickness.

Shear Strength
When bending is produced by transverse forces, as in three-point bending, the sandwich beam experiences shearing in addition to bending. When the faces are thin, the shear stress is uniform over the depth of the core and is given by:

$$\tau = \frac{Q}{bd}$$

where τ = the shear stress and Q = the shear force.

Hence if τ^* is the ultimate shear stress of the core material, then the shear strength of the beam is $Q_u = bd\,\tau^*$. Expressed in N/mm width:

$$\frac{Q_u}{b} = d\tau^*$$

The shear strength of the sandwich beam is almost independent of the faces and is mainly a function of the core shear strength and thickness.

Flexural Rigidity

The flexural rigidity (D) is given by:

$$D = \frac{bE_f t d^2}{2}$$

where E_f = the modulus of the face material. $E_f t$ is the tensile stiffness of the faces per unit width (the minimum values for various materials are given in Table 1.16).

Flexural rigidity per mm width is plotted against reinforcement weight, for four different core thicknesses, c, for sandwich constructions made with GFRP faces using CSM and WR in Fig. 1.27 and 1.28 respectively. As in the case of bend strength, the R : F ratios for CSM and WR were chosen as 2.33 : 1 and 1 : 1 respectively. The R : F ratio of the faces has rather more influence on the flexural rigidity of the sandwich construction than on its bend moment at failure, but the effect is still small compared to that of core thickness as a result of the $d^2[(c+t)^2]$ term.

Shear Stiffness

Shear stiffness of a sandwich beam is given by:

$$N = \frac{bG_c d^2}{c}$$

where G_c = the shear modulus of the core material.

For thin faces $d \simeq c$: hence the shear stiffness is approximately proportional to the core thickness, i.e. the faces do not then contribute significantly to the shear stiffness.

Deflection of Simply Supported Beam with Midpoint Loading

The deflection is the sum of two parts, the first due to bending and the second to shearing:

$$\Delta = \frac{WL^3}{48D} + \frac{WL}{4AG}$$

where W = midpoint load, L = span and $A = \frac{bd^2}{c}$

The bending and shear stiffnesses, D and N, can be calculated as shown earlier, and used in the deflection equation to show the relative contributions to bending and shearing. For example, in a three-point bend test in a sandwich construction at 16 : 1 span to depth ratio (the usual ratio for solid plastic materials), the deflections due to bending and shearing can be calculated as follows (assuming skins of one layer of 450 g/m² CSM at R : F = 2.33 : 1 and 10 mm core thickness of PU foam with a shear modulus of 5 N/mm² and beam width of 50 mm).

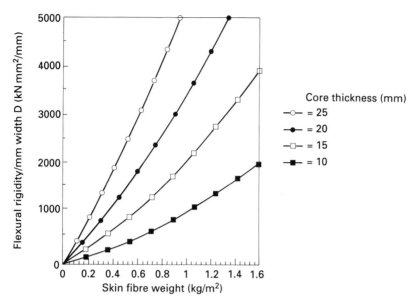

1.27 The effect of CSM skin construction (at R : F = 2.33 : 1) and core thickness on flexural rigidity for balanced double skinned sandwich laminates.

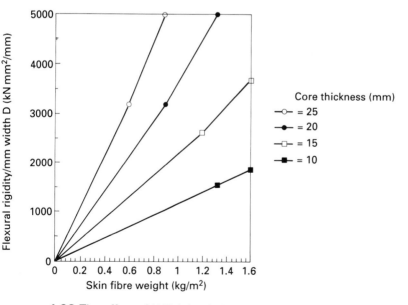

1.28 The effect of WR (glass) skin construction (at R : F = 1 : 1) and core thickness on flexural rigidity for balanced double skinned sandwich laminates.

From Fig. 1.10,

face thickness = 0.2 (glass) + 0.8 (resin) = 1.0 mm

From Table 1.16,

extensibility for CSM = 14 000 N/mm²/kg/m² of glass

Hence,

$$Et = 14\ 000 \times 0.45 \text{ (wt of glass in kg/m}^2)$$
$$= 6\ 300 \text{ N/mm}$$

Since $t=1$ mm, $c=10$ mm, $d=c+t=11$ mm, $h=c+2t=12$ mm, $b=50$ mm and $G_c=5$ N/mm²,

$$D = \frac{bE_f t d^2}{2} \quad = 19.1 \times 10^6 \text{Nmm}^2$$

and

$$AG = \frac{bG_c d^2}{c} = 3025 \text{ N}$$

At 16 :1 span : depth ratio, span $L = 16h = 192$ mm, and with a load $W = 250$ N applied at the midpoint,

$$\text{bend deflection} = \frac{WL^3}{48D} = 1.93 \text{ mm}$$

and

$$\text{shear deflection} = \frac{WL}{4AG} = 3.97 \text{ mm}$$

i.e. the shear deflection is approximately double the bending deflection. Hence, if the shear deflection is ignored, or not recognised, in a three-point bend test at 16 : 1 span to depth ratio, a false value of flexural rigidity, only about a third of the correct value, will be obtained.

At a span to depth of ratio of 80 : 1, i.e. $L = 960$ mm, the shear deflection is only 8% of the bend deflection.

Since most sandwich constructions are fairly thick, the length of specimens and test rigs required for three-point bend testing must be large, often one metre or more if shear effects are to be avoided.

The theme in this section has been to strike a balance between data measurement and data prediction and as databases grow, with more reliable data available, so the initial screening for a material to meet a given structural requirement need only be theoretical, thus reducing the level of testing required to prove the cost-performance criteria of composites for the application. However, final proof of the performance of the chosen structural laminate should always involve mechanical testing to validate the predictions.

Table 1.19 Relative and specific properties of various fibre reinforced laminates

Relative or specific property	CSM	Glass WR-CSM	Poly-aramid WR	Glass WR	Poly-aramid WR	UD–CSM	UD poly-aramid–CSM
Relative weights of FRP panel (designed for a pressure of 2 kg/cm² in membrane at ultimate stress)	1	0.9	0.3	0.57	—	—	—
Relative weights of FRP panel under-pressure and membrane for a given deflection	1	0.91	0.67	0.85	—	—	—
Relative weights for equal flexural strength	1	0.88	0.75	0.91	0.78	—	—
Relative weights for equal stiffness in bend	1	0.88	0.66	0.83	0.56	—	—
Specific compressive moduli	560	775	1370	1120	1260	1220	1950
Specific tensile moduli	560	760	1385	980	2010	1350	2250
Specific compressive strength	10.5	9.5	11	12	9	15.5	10
Specific tensile strength	7.5	12	28	15.5	32.5	29.5	44.5

It must be remembered that raw material cost should not be considered in isolation from processing cost, since the benefits of expensive higher performance materials can often be realised at little or no premium because of favourable processing costs. Comparison between various fibre constructions and relative and specific properties are given in Table 1.19.

It is clear that the properties of composite materials are extremely diverse and this section has simply served as an introduction to their performance potential. Each case obviously requires careful analysis to meet the design requirements of load-bearing structures.

1.6 Health and Safety

No discussion of materials used in large quantities under workshop conditions would be complete without discussion of the safe use and handling of the materials. Composite materials are very different from other structural materials in that as the component is being fabricated, so too is the material being made at the same time, usually from a liquid base.

Hence, not only do the normal precautions need to be taken as would be the case when handling any material (e.g. against inhalation, ingestion of dust, eye injury, injury from process techniques), but also precautions must be taken against handling liquid chemicals which may be inflammable, corrosive, hazardous from inhalation, ingestion, skin contact, etc.

In general, most materials used in the composite industry have low risk to health status, but that should be no excuse for complacency, since the best advice is to avoid unnecessary contact with any chemical by using appropriate and recommended safety precautions to reduce all risks. Further, to ensure that composites do not become commercially unviable for some applications, through bad practice, resulting in an undeserved unsafe label, care should be taken not to attract strict unnecessary legislation.

1.6.1 The Simple Rules

Do:

Store and handle raw materials in accordance with the supplier's instructions and legal requirements.

Be aware of health and safety hazards associated with the process.

Ensure that catalyst and accelerators are never stored together, or with resin.

Always have an inert, absorbent material available in case of spillage.

Provide and use the appropriate protective clothing and cleaning materials.

Protect against the toxic and harmful effects of the raw materials by providing extraction and dust control.

Ensure adequate ventilation and fume control.

Ensure good standards of housekeeping.

Ensure that if respiratory protective equiment is used, that it is suitable for the purpose.

Use materials with low emissions wherever possible.

Never:

Directly mix catalyst and accelerator.

Smoke in stores and working areas.

Use sawdust or combustible materials to absorb spillages.

Use solvents for cleaning hands.

Allow waste to accumulate.

Table 1.20 Physical properties, occupation exposure limits and health hazards for polyester resins based on various monomers

Property	Monomer base			
	Styrene	Methyl methacrylate	Vinyl toluene	α-Methyl styrene
SG	0.910	0.936	0.916	0.909
Flashpoint (°C)	31	11	60	54
Boiling point (°C)	145	101	172	165
Auto ignition temperature (°C)	490	430	575	574
Vapour pressure (mm/Hg at 20 °C)	6.1	2.8	1.1	1.9
Lower explosive limit (%)	1.1	2.1	13.8	1.9
Upper explosive limit (%)	8.0	12.5	30.0	6.1
OEL	8 h TWA = 100 ppm Short term exposure = 250 ppm	100 ppm 125 ppm	100 ppm 150 ppm	100 ppm 100 ppm
Odour threshold	0.25 ppm	—	—	—
Acute effects of inhalation	Little or none less than 210 ppm; appreciable irritation more than 500 ppm			
Ingestion	Harmful effects >4 g/kg of body weight	Harmful effects >9 g/kg of body weight	As for styrene	As for styrene
Skin splashes	Degreases skin, causing irritation. Frequent prolonged contact can cause dermatitis.		Excessive prolonged contact may produce chemical burn.	
Eye splashes	Mild to severe irritation which can last several hours.			

1.6.2 Resins

1.6.2.1 Unsaturated Polyester Resins

The Base Resins

The prime risks come from the monomer, usually styrene and/or other chemicals used to produce the resin solution. The volatile nature of the monomer makes resin solution flammable, and care must be taken not to expose operators to unacceptable levels of the vapour during moulding processes. Table 1.20 gives safety-related properties of polyester resins with various monomer types. The occupational exposure limits (OEL) prescribe the levels of exposure based on an average exposure period of eight hours, above which employees should not be exposed – usually expressed as the time weighted average (TWA).

Catalysts
Peroxides are severe irritants and corrosive to moist tissue in the eyes, nose, throat and airways to the lungs. Irreversible damage can be caused to the eyes by prolonged contact. Some catalysts cause severe skin irritation.

Accelerators
Cobalt based accelerators are solutions of organic cobalt compounds in styrene. The solutions are flammable and will irritate the eyes, skin and lungs at concentrations above the OELs of the constituents.

Amine based accelerators are solutions of tertiary amines usually in styrene at strengths of up to 25%. Amines are toxic and even the low concentration accelerators should be treated as such. They are toxic if swallowed, inhaled or allowed to remain in contact with the skin for any appreciable time. The fumes from burning accelerators also contain toxic materials.

Cleaning Solvents
The most common cleaning solvents are acetone, styrene and methylene chloride. All are harmful and precautions must be taken to reduce exposure to a minimum. In no circumstances should the relevant occupational exposure level be exceeded. Never use equipment cleaning solvents to clean the skin.

Table 1.21 shows the properties and OELs of cleaning solvents.

1.6.2.2 Epoxy Resins

The Base Resin
The unmodified liquid diglycidyl ether of bisphenol-A resins are described as being mild to moderate primary skin irritants whose irritation potential is increased by prolonged skin contact. The diluents present render some resins mild to strong sensitisers, depending upon the chemical nature of the diluent. Other solution grades can be mild to moderate skin irritants and possible skin sensitisers.

Fully cured epoxy resins, like polyester resins, are practically non-toxic, non-irritant and non-sensitising to the skin.

Dust produced during the machining of epoxy resins and composites can present a health hazard and inhalation of dust should be avoided.

The Curing Agents
For all classes of curing agents used with epoxy resins, protective clothing should be worn, and ventilation is essential. Goggles should also be worn to prevent the possibility of acute eye irritation as a result of eye contact with the curing agents.

Aliphatic amines are alkaline, caustic materials which can cause

Table 1.21 Properties and occupation exposure levels of cleaning solvents

Property	Methylene chloride	Acetone
SG	1.33	0.79
Flashpoint (°C)	Non-flammable but can decompose under moist heat to HCl, CO_2, CO and possibly $COCl_2$	−17.8
Boiling point (°C)	40	56.2
Auto ignition temperature (°C)	615	465
Vapour pressure (mg/Hg at 24 °C)	400	226.3
Lower explosive limit (%)	Not applicable	2.1
Upper explosive limit (%)	Not applicable	13.0
Control limit	8 h TWA = 100 ppm Short term = 250 ppm	100 ppm 1250 ppm
Inhalation	Exposure above control limit results in an increase in blood CO level	Narcosis above the recommended level of exposure
Ingestion	Chemical burning of mouth	Corrosive if ingested; could perforate stomach
Skin splashes	Chemical burns	Irritation and dermatitis
Eye splashes	Severe chemical burning and permanent damage	Irritation
Chronic effects	Long term effects on the heart. Damage to kidneys and liver as a result of ingestion	—

burns and severe tissue damage to skin, mucous membranes and eyes. All contact should be avoided.

The solid *aromatic amines* are less caustic, less irritating and less sensitising than the aliphatic amines. Diaminodiphenylmethane is a toxic chemical known to cause liver damage in humans. It can be absorbed through the skin, so all contact should be avoided.

Cycloaliphatic amines vary in their irritation and sensitising effect, depending upon type and molecular weight. They are extremely irritating to the eyes.

Polyamide curing agents are skin irritants of varying sensitivity, but are generally non-sensitising to the skin. They are, however, extremely irritating to the eyes.

1.6.2.3 Phenolic Resins

The *base resin* should be handled and used in well-ventilated areas, preferably in enclosed systems. Phenolic resins may contain appreciable quantities of free phenols and/or free aldehydes. Phenol, substituted phenols and formaldehyde are toxic by inhalation, ingestion, skin contact and mucous membrane contact.

Solid resins present little or no vapour hazards at normal ambient temperatures, but inhalation of dust and powder must be avoided. Vapour inhalation from water-based phenolic resins can cause respiratory problems due to the presence of unreacted phenol and formaldehyde.

Ingestion of phenolic resin solution results in severe irritation of the alimentary tract and can prove fatal.

Phenolic resins can present a greater dermatitis risk than some other resin types. Skin contact should be avoided, but in such occurrences the resin should be washed off with water before it dries. Splashes of water-soluble resin in the eye can cause permanent blindness if not removed very rapidly with copious quantities of water.

During thermal decomposition, phenolic resin fumes contain phenol, cresols or alkyl phenols, aldehydes – notably formaldehyde – and other monomeric substances.

The *curing agents*: the common acid hardeners are based on aromatic sulphonic acids and phosphoric acid. These hardeners react violently with resins, bases and peroxides, and give off hydrogen gas if they come into contact with base metals such as mild steel. Acid hardeners are also corrosive to skin and clothing.

Sulphonic acid catalysts can cause dermatitis, skin sensitisation or asthma. Severe eye damage can occur from contact, and prolonged inhalation can cause chronic bronchitis.

1.6.2.4 Furane Resins

Furane resins cause eye irritation and prolonged skin contact may cause irritation and should be avoided. The catalysts used with furane resins are generally strongly acidic, corrosive to the skin and severe eye irritants.

1.6.3 Reinforcements

The most common reinforcements – glass, carbon and polyaramid fibres – are not generally considered to be harmful. However, there is still debate about possible hazards associated with the dust of particularly, glass fibre. Currently a control limit of 10 mg/m^3 has been adopted for total dust and 5 mg/m^3 for respirable dust from man-made mineral fibres.

Care should be taken when handling reinforcement, to ensure that these levels are not exceeded.

1.6.4 Fillers and Pigments

Resins are generally supplied with fillers and pigments predispersed. Dispersion of fillers and dusty powders by users of resin is not recommended unless the appropriate handling facilities are used. Generally, fabricators need only concern themselves with filler toxicity as a result of dust production from resin-containing fillers during machining operations on filled fibre reinforced laminates. In these instances, suitable dust extraction equipment will reduce any possible hazard.

1.6.5 Foam Cores

In situ foaming of polyurethane should be carried out with care, since such foams contain isocyanates – chemicals which are known to cause lung problems at low concentrations and sensitisation by excessive or continued over exposure, such that concentrations well below the control limit (0.02 mg/m^3 of isocyanate) trigger an asthmatic reaction. Isocyanates may also irritate the skin and eyes.

Other foam core, such as PVC foam supplied in slab-stock, are considered to be no toxic risk other than from a dust hazard as a result of machining; this should be handled as any other dust hazard would be with fibre reinforced plastic.

Part II

Techniques for manufacture of composites

2 Manufacturing Processes

This chapter describes the methods of making articles in fibre reinforced composites. The materials will be restricted to *thermosetting* resins reinforced with *glass, carbon* or *aramid* fibres.

2.1 Introduction

Thermosetting resins are among the simplest of plastic materials to process. All that is necessary is to mix in the activator, pour into a mould, leave to set and then remove from the mould. An exact reproduction of the mould contours is produced, without any need for heat or pressure. The mould can be made from almost any non-porous material – metal, wood, plastic, even plaster or cardboard if the surface is sealed.

If a higher production rate is required, a heated mould can be used. Heat speeds up the curing process, produces a more complete cure and allows the use of activators which give a longer pot life.

The main problems associated with moulding unreinforced thermosets are:

1 The possibility of air bubbles being trapped in the mould.
2 Design of the mould so that the component can be extracted from it.
3 Shrinkage. Thermosetting resins can reduce in volume by up to 8% during the setting and curing process. This is increased if there is a temperature rise during curing as the thermal contraction adds to the molecular shrinkage. It is important to remember that the hardening reaction produces heat; consequently large masses of resin can become hot even if they are being 'cold-cured'. Rapid and uneven shrinkage generates internal stresses in the cured part and can cause cracking.

2.1.1 The Five Process Elements

It can be seen from the above description that the conversion of unreinforced thermosetting resins involves three distinct operations:

1   Mixing resin and activator.
2   Dispensing resin into the mould.
3   Curing.

The introduction of reinforcement in any form complicates the conversion process by adding two further operations:

4   Positioning reinforcement.
5   Impregnating reinforcement with resin.

All fibre reinforced polymer composite manufacturing processes contain these five elements. Methods of executing the individual elements can be thought of as 'techniques'. Each manufacturing process can therefore be considered as an assembly of techniques for executing the five elements. This is an important point because it indicates the large number of processing arrangements which are possible with composites by using different combinations of techniques. You will see that many of the processes described on the following pages differ in only one or two of the five elements. To avoid repetition individual techniques are only described in detail once. Cross-referencing may therefore be necessary to obtain a full description of any particular process.

2.2 Open Mould Processes

2.2.1 Hand Laminating

Although messy, labour-intensive and difficult to control, the hand laminating process is still widely used because of its inherent flexibility and the low outlay in moulds and equipment. The various stages of this technique are shown in Fig. 2.1.

In its basic form resin is mixed by hand in a bucket and applied by brush or mohair roller to the mould. Normally a neat resin layer called the 'gel coat' is applied first and allowed to gel before proceeding with reinforced layers. Reinforcement is usually cut to size prior to mixing the resin. Alternate layers of resin and reinforcement are then applied to the mould and a ribbed metal roller used to consolidate the laminate and fully impregnate the reinforcement, while at the same time forcing all air out. Large or thick components are usually made in several stages, allowing the resin to gel after each stage before proceeding with the next. In the same way stiffening ribs, core materials and metal inserts can be fitted during the build-up of the laminate and encapsulated by later layers.

A data sheet for hand laminating is given in Table 2.1.

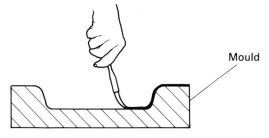

Mould

(a) Apply gel-coat with brush or soft roller
Allow to gel.

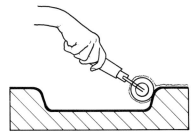

(b) Apply laminating resin with brush or soft roller

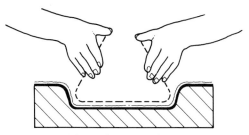

(c) Cut and fit reinforcement layer

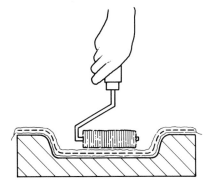

(d) Consolidate with ribbed roller.
Repeat (b), (c) and (d) until required
build-up is achieved

2.1 Hand laminating.

Table 2.1 Composites process data sheet: hand laminating

Composites process data sheet	
Name: Hand laminating	**Type:** Open mould

Other names: contact moulding
              hand lay-up

Abbreviations: HLU

See also:

Process parameters
Fibre volume fraction: 0. 13–0.5
Size range: 0.25–2000 m²
Processing pressure: ambient
Processing temperature: ambient
Production rate: 1–4/day/mould

Materials
Reinforcement: most types can be used.
Resins: polyester, epoxy, vinyl ester, phenolic, etc.
Fillers: up to 10% by volume
Core materials: PU or PVC foam
                Balsa
                Honeycomb

Cost factors

Production cost:	High
Material cost:	Moderate
Tooling cost:	Low
Equipment cost:	Very low

Tolerances

Overall dimensions:	1.0–5.0 mm
Thickness:	1.0–3.0 mm
Moulded detail:	0.5–1.0 mm
	One surface only

Design Details

Minimum radii:	1.0 mm
Ribs:	Yes
Bosses:	Difficult
Holes:	Can be moulded-in
Thickness changes:	Yes

Part geometry
Any thin-walled shape which allows access for operator. Complex detail difficult to mould.

Advantages
Low capital outlay
Secondary bonding
No size limit
Flexibility

Disadvantages
Operator-dependent
Labour-intensive
Low production rate
Poor weight and thickness control
Only one moulded face

2.2.2 Saturation

Saturation is the first stage in the mechanisation of open mould lay-up. Instead of the resin being mixed in a bucket and applied by brush or roller, it is mechanically mixed and sprayed on to the mould.

Most modern spray guns use an air drive pump which draws resin straight from the drum. Linked to it is a catalyst pump. The pumps deliver resin and catalyst to a hand-held, trigger-operated gun. Catalyst and resin are mixed in the gun and then ejected under pressure. Spray guns of this type can only operate with low viscosity resins. Some can be used for both gel coat and lay-up resins.

Saturation offers improved control over resin mixing, but resin distribution control is still in the hands of the operator. Fibre volume content and thickness variations can therefore be expected to be as bad as for hand laminating.

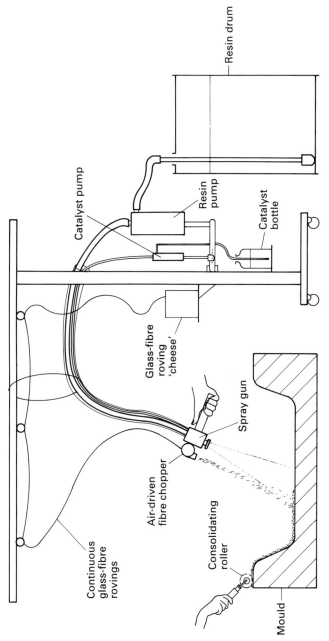

2.2 Spray-up.

2.2.3 Spray-up

Spray-up is a process in which both resin and fibre are sprayed on to the mould simultaneously. An air-driven chopper unit is mounted on a resin spray gun. The chopper devours strands of continuous reinforcement, cuts them and spits them out in short lengths. The chopped fibre is carried to the mould by the resin stream so that reinforcement and resin are dispensed and positioned simultaneously. Chopper and spray gun output can be matched to give the required fibre content. Consolidation is by hand rolling as in hand laminating, but is easier as the reinforcement contains no binder and the fibre content is more uniform. The technique is demonstrated in Fig. 2.2.

Spray-up is a highly productive process but control of fibre distribution and thickness are entirely in the hands of the operator and it is by its nature restricted to random chopped reinforcement. This makes the process unsuitable for critical applications, but it is popular for high volume, non-critical products.

A data sheet for spray-up is given in Table 2.2.

Table 2.2 Composites process data sheet: spray-up

Composites process data sheet	
Name: Spray-up	**Type:** Open mould
Other names: Spray lay-up	**Process parameters** Fibre volume fraction: 0.13–0.21
Abbreviations: None	Size range: 2.0–100 m²
	Processing pressure: ambient
See also: auto spray-up spray winding	Processing temperature: ambient Production rate: 1–4/day/mould
Materials Reinforcement: chopped roving only Resins: polyester, vinyl ester Fillers: not recommended Core materials: PU or PVC foam Balsa	**Cost factors** Production cost: Low Material cost: Low Tooling cost: Low Equipment cost: Moderate
Tolerances Overall dimensions: 1.0–3.0 mm Thickness: 2.0–3.0 mm Moulded detail: not recommended	**Design details** Minimum radii: 6.0 mm Ribs: Yes, foam-filled Bosses: No Holes: below 250 mm: Drill or cut above 250 mm: Mould or cut

Part geometry
Best suited to simple open shapes with no
  fine detail
	Thickness changes: difficult because poor
	  thickness control

Advantages Low material cost High production rate Low tooling cost Large parts	**Disadvantages** Very operator-dependent Very poor thickness control Only one moulded face Random reinforcement only

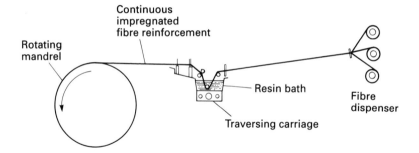

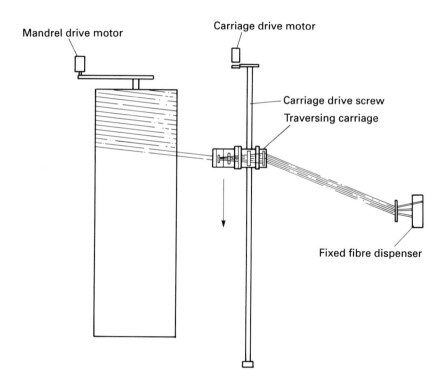

(a)

(b)

2.3 Filament winding: (a) sectional elevation; (b) plan view.

2.2.4 Auto spray-up

The major disadvantage of the hand-held spray-up process can be overcome if the gun orientation and speed are under the control of a machine, which can be designed or programmed to deposit the materials as required. This is straightforward for simple shapes, and dedicated machines are available for the automatic spray-up of flat rectangular panels, cylinders and dished ends. Such machines usually incorporate automatic rolling as well as spraying.

More complex shapes demand some form of numerical control. Robot manipulators have been used with some success but are still not ideal if used in 'teach' mode as this only reproduces the best that a human operator can achieve.

2.2.5 Filament winding

Figure 2.3 shows the filament winding process that is employed for the production of simple hollow shapes and is particularly suitable for pressure vessels. The component is moulded on a male former, giving a moulded surface on the inside. The former is mounted on a shaft which is fitted to a lathe so that it can rotate. A resin bath is mounted on the lathe-traversing head. Continuous fibre strands are fed through the resin bath from fixed positions alongside the lathe. The fibres emerge from the bath via nip rollers (to remove excess resin) and then pass through a vertical comb before being wrapped around the rotating former.

By changing the relative speeds of mandrel and traverse the winding angle can be controlled. The fibre orientation and thickness can therefore be varied to give optimum performance. The tension applied to the reinforcement can also be controlled to achieve optimum fibre content and good consolidation of the laminate.

Filament winding has been used for many years to produce tanks, pressure vessels, rocket motors, gas bottles, etc., and currently accounts for most of the composite pipe manufactured.

A data sheet for filament winding is given in Table 2.3.

2.2.6 Spray winding

Spray winding is a combination of auto spray-up and filament winding. Layers of random chopped fibre are interleaved with wound-on continuous fibre strands to build up thickness on a rotating former. Resin is applied by spray gun with the chopped fibres as in auto spray-up. It is more cost-effective for low pressure applications than a totally filament-wound laminate.

Table 2.3 Composites process data sheet: filament winding

Composites process data sheet

Name: Filament winding | **Type:** Open mould

Other names:	hoop winding spiral winding	**Process parameters** Fibre volume fraction: 0.55–0.7 Size range: 0.1–100 m^2	
Abbreviations:	FW	Processing pressure: ambient Processing temperature: ambient	
See also:	spray winding, continuous filament winding	Production rate: 1–5/day/mould	

Materials
Reinforcement: continuous rovings only
Resins: polyester, epoxy, vinyl ester
Fillers: can be used
Core materials: foam or Balsa

Cost factors
Production cost: Low
Material cost: Low
Tooling cost: Moderate
Equipment cost: Moderate to high

Tolerances
Overall dimensions: 1.0–2.0 mm
Thickness: 0.5–1.0 mm
Moulded detail: 0.5–1.0 mm

Design details
Minimum radii: 1.0 mm
Ribs: hoop only
Bosses: no
Holes: normally drilled or cut
Thickness changes: yes, within winding
 pattern

Part geometry
Must be a solid of revolution

Advantages
Excellent mechanical properties
High production rate
Good control of fibre orientation
Good thickness control
Good fibre content control
Good internal finish

Disadvantages
Limited range of shapes
Limited number of practical winding
 patterns

2.2.7 Centrifugal casting

Centrifugal casting, shown diagrammatically in Fig. 2.4, is used to produce hollow articles, and gives a moulded surface on the outside.

Resin and reinforcement are placed inside a cylindrical mould which is rotated at high speed. The centrifugal acceleration forces the materials against the mould surface, expelling air and consolidating the laminate. The reinforcement, being more dense, tends to move to the outer surface, while the inner surface becomes resin-rich. For shapes other than cylinders this effect tends to reduce thickness at points closer to the axis of rotation and to increase it at points further away.

Where random reinforcement only is required, a traversing spray-up gun is an effective way of applying resin and reinforcement together. Where a more complex construction is needed, the dry reinforcement pack can be wrapped around a mandrel and loaded into the mould. The mandrel is then removed and the mould rotated to force the layers

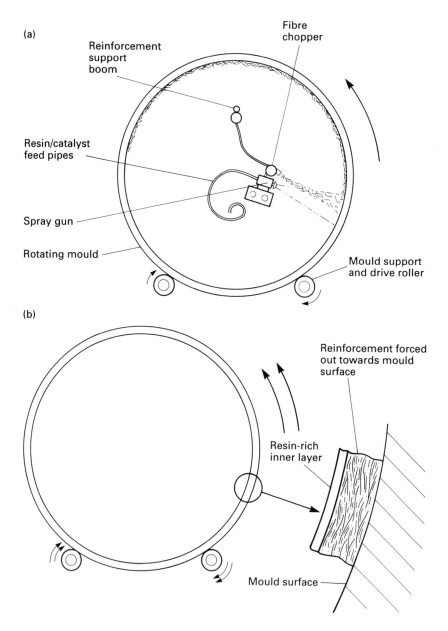

2.4 Centrifugal casting: (a) random reinforcement and resin are sprayed on to slowly rotating mould by traversing gun; (b) mould rotates rapidly to consolidate laminate.

out into their proper position. Resin is then deposited by a traversing gun while the mould is rotating.

The process is used to make tanks, pipes and poles for street lighting, telephone cables and flags.

A data sheet for centrifugal casting is given in Table 2.4.

Table 2.4 Composites process data sheet: centrifugal casting

Composites process data sheet	
Name: Centrifugal casting	**Type:** Open mould

Other names: none	**Process parameters** Fibre volume fraction: 0.2–0.6
Abbreviations: none	Size range: 0.5–100 m^2 Processing pressure: ambient
See also:	Processing temperature: 40–60 °C Production rate: 4–8/shift/mould
Materials	
Reinforcement: mats or cloths	**Cost factors**	
Resins: polyester, vinyl ester	Production cost: Low
Fillers: can be used	Material cost: Moderate
Core materials: no	Tooling cost: High
	Equipment cost: Moderate to high
Tolerances	**Design details**
Overall dimensions: 1.0–3.0 mm	Minimum radii: 10 mm
Thickness: 0.5–1.0 mm	Ribs: no
Moulded detail: 0.5–1.0 mm	Bosses: no
Max. depth 1.0 mm	Holes: drill or cut
	Thickness changes: gradual
Part geometry	
Must be a solid of revolution	
Advantages	**Disadvantages**
High production rate	Limited to parallel or tapered cylinders
Good fibre content control	Non-moulded inner surface
Good thickness control	
Very good consolidation	
Resin-rich inner surface	

2.3 Closed mould processes

2.3.1 Vacuum bag

Vacuum bag moulding, shown in Fig. 2.5, is the simplest form of the closed mould process. Reinforcement and resin are applied by hand laminating to a simple open mould; a release film is then laid over the laminate, followed by a rubber bag which is clamped to the edge of the mould. The space between the bag and the mould is evacuated so that atmospheric pressure is applied over the surface of the laminate. Additional rolling on the outside of the bag may be necessary to achieve complete consolidation.

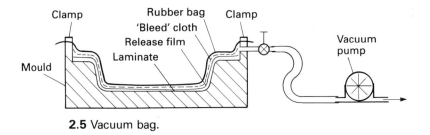

2.5 Vacuum bag.

Table 2.5 Composites process data sheet: vacuum bag

Composites process data sheet	
Name: Vacuum bag	**Type:** Closed mould

Other names: none	**Process parameters**
	Fibre volume fraction: 0.15–0.6
Abbreviations: none	Size range: 0.5–20 m^2
	Processing pressure: 1 bar
See also: autoclave	Processing temperature: ambient
	Production rate: 1–3/shift/mould
Materials	
Reinforcement: most types	**Cost factors**
Resins: most types	Production cost: High
Fillers: not recommended	Material cost: Moderate to high
Core materials: Foam	Tooling cost: Low
Balsa	Equipment cost: Low
Honeycomb	
Tolerances	**Design details**
Overall dimensions: 1.0–3.0 mm	Minimum radii: 1.0 mm
Thickness: 1.0–2.0 mm	Ribs: yes
Moulded detail: 0.5–1.0 mm	Bosses: difficult
	Holes: can be moulded in
Part geometry	Thickness changes: yes
Any thin-walled shape which allows access for operator. Complex detail difficult to mould	
Advantages	**Disadvantages**
Low capital outlay	Labour-intensive
Low cost tooling	Low production rate
Large components	Only one accurate surface
Well suited to making sandwich panels	

Vacuum bagging produces a fairer finish than hand laminating, but atmospheric pressure is not sufficient to distribute resin, impregnate and consolidate, so that a large part of the hand laminating process remains.

Vacuum bag is a very effective method for bonding sandwich laminates together.

A data sheet for vacuum bag is given in Table 2.5.

2.3.2 Pressure bag

Pressure bag moulding is a similar process to vacuum bag and allows the use of higher pressures than atmospheric (up to 3.5 bar), producing better consolidation and higher fibre content. Faster and/or better curing can also be achieved by using heated air or steam in the bag. However, the mould needs to be much more robust as it has to withstand the loads generated by the applied pressure.

Figure 2.6 illustrates the process diagrammatically.

The use of a proprietary bag press makes this process reasonably productive. It is normally used for the production of high quality components made with preimpregnated reinforcement (prepreg). In a prepreg the resin is in a partially cured state (B-stage) which gives it a putty-like consistency. Under the action of heat and pressure the prepreg forms to the shape of the mould and then sets and cures.

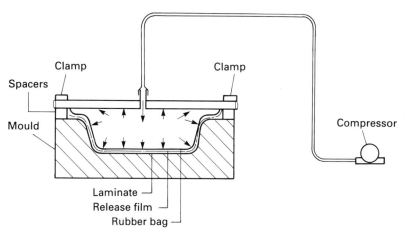

2.6 Pressure bag.

2.3.3 Autoclave

Autoclaving is a combination of vacuum and pressure bag moulding, shown in Fig. 2.7, which produces the very highest quality components. It is nowadays used almost exclusively with prepreg materials.

The process uses a vacuum bag assembly inside a heated and pressurised vessel. Layers of prepreg material are laid on the mould to make up the full thickness. Bleed cloth, release film and vacuum bag are placed over the prepreg and sealed on the mould using partial vacuum. The mould and its contents are then loaded into the autoclave, which is closed and sealed. Inside the autoclave the laminate is subjected to vacuum, pressure and heat simultaneously. This ensures that all air is extracted from the laminate and full consolidation and cure are achieved.

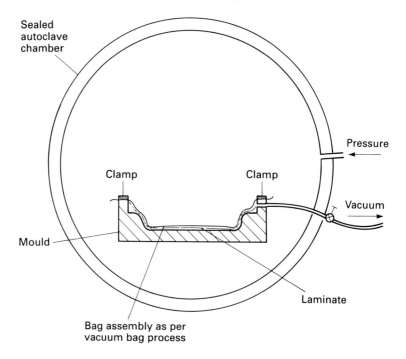

2.7 Autoclave.

An important advantage of the process is that moulds are not subjected to large forces and can therefore be of reasonably light construction.

A data sheet for autoclave is given in Table 2.6.

2.3.4 Leaky mould

Leaky mould is the simplest form of matched mould process. Male and female moulds are used, which when clamped together form a cavity the exact shape of the finished component.

Resin and reinforcement are laid in the hollow (female) mould by hand laminating. The upper (male) half of the mould is placed over the female and squeezed down onto stops using a press or G-clamps. The two halves of the mould are made to be a loose fit so that excess resin is ejected as closure is completed. When curing is complete the clamps are removed, the mould split and the component extracted.

Matched moulds produce components with accurate dimensions and a good quality finish on both sides.

2.3.5 Cold press

To gain the full benefit of pressure moulding it is necessary to use rigid moulds mounted in a hydraulic press which is capable of exerting a

Table 2.6 Composites process data sheet: autoclave

Composites process data sheet	
Name: Autoclave	**Type:** Closed mould

Other names: none	**Process parameters**
	Fibre volume fraction: 0.35–0.7
Abbreviations: none	Size range: 0.25–5.0 m²
	Processing pressure: up to 10 bar
See also: vacuum bag	Processing temperature: 140 °C
	Production rate: 1/shift/mould
Materials	
Reinforcement: undirectional and woven	**Cost factors**
prepregs	Production cost: High
Resins: Epoxy	Material cost: High
Fillers: no	Tooling cost: Moderate
Core materials: Honeycomb	Equipment cost: High
Balsa	
Tolerances	**Design details**
Overall dimensions: 0.5–1.0 mm	Minimum radii: 3.0 mm
Thickness: 0.1–0.5 mm	Ribs: top-hat with core
Moulded detail: no	Bosses: no
	Holes: drill or cut
Part geometry	Thickness changes: yes
Best with simple shallow shapes	
Advantages	**Disadvantages**
Very high quality	Labour-intensive
High fibre content	Slow
Low void content	High capital investment
Controlled cure	

pressure of at least 2 bar. With this arrangement it is possible to use the compression operation to distribute the resin and impregnate the reinforcement with it, purging air from the mould at the same time. This requires a properly matched set of moulds with positive location on closing. Figure 2.8 shows the process diagrammatically.

The reinforcement pack is assembled dry and loaded on to the mould. The required quantity of resin is mixed and poured on to the reinforcement. The mould is then closed and full pressure applied. When the resin has hardened, the press is opened and the part removed.

Cold press moulding produces accurate components at a reasonable rate with modest tooling costs.

A data sheet for cold press moulding is given in Table 2.7.

2.3.6 Hot Press

The rate of production can be increased dramatically by applying heat to the mould surface to accelerate the cure process. To achieve the highest output the mould needs to be at 140 °C and this necessitates the use of metal moulds.

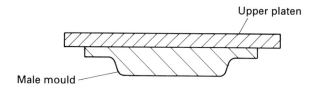

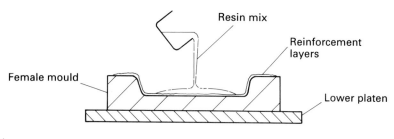

(a)

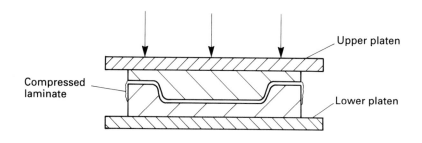

(b)

2.8 Cold press: (a) press open – loading resin; (b) press closed – pressure applied.

Hot press moulding can be carried out in the same way as cold press, using liquid resin. However, it is nowadays normal to use prepreg reinforcements, either in the form of continuous fibre prepregs or sheet and dough moulding compounds, which are known as SMC and DMC (or BMC). These compounds contain chopped fibre reinforcement and are formulated so that both fibre and resin are able to flow under the action of heat and pressure. This allows complex details to be moulded from a simple material pack.

SMC and DMC cycle times are normally between two and ten minutes; moulding pressures required are much higher than with cold press and quantities of less than 10 000 are unlikely to be economic because of the high cost of tooling. Very high accuracy and stability are

Table 2.7 Composites process data sheet: cold press moulding

Composites process data sheet

| **Name:** Cold press moulding | **Type:** Closed mould |

Other names: none	**Process parameters**
	Fibre volume fraction: 0.15–0.25
Abbreviations: CPM	Size range: 0.25–5.0 m²
	Processing pressure: 2–5 bar
See also: resin injection	Processing temperature: 20–50 °C
	Production rate: 10–30/shift/mould

Materials
Reinforcement: continuous strand
                 mat/woven cloth
Resins: polyester
Fillers: up to 20% by volume
Core materials: no

Cost factors
Production cost: Low
Material cost: Moderate
Tooling cost: Moderate
Equipment cost: Moderate

Tolerances
Overall dimensions: 0.25–1.0 mm
Thickness: 0.25–0.5 mm
Moulded detail: 0.25–0.5 mm

Design details
Minimum radii: 3 mm
Ribs: no
Bosses: no
Holes: cut or drill
Thickness changes: yes

Part geometry
Best with simple shallow shapes
Must allow mould separation
Minimum draw angle 2°

Advantages
Good surface both sides
Good production rate
Accurate dimensions

Disadvantages
Limited by press size
Low fibre content

---

possible, but variability in mechanical properties can be a problem.
A data sheet for hot press moulding is given in Table 2.8.

2.3.7. Resin injection

As with cold press moulding the reinforcement pack is made up and loaded into the mould. In this process, however, the mould is closed onto the dry pack before the resin is introduced. Resin is mixed and pumped into the mould by an air-driven dispensing machine through one or more injection points. Fill time can be between one and ten minutes depending on component size and fibre content. The process is illustrated in Fig. 2.9.

Injection pressure can be up to 2 bar, so mould construction needs to be substantial to resist the loads generated without undue distortion. Mould-handling equipment is therefore required for all but the smallest components.

In its basic form resin injection is limited to random reinforcement

Table 2.8 Composites process data sheet: hot press moulding

Composites process data sheet

Name: Hot press moulding | **Type:** Closed mould

Other names: Compression moulding

Abbreviations: none

See also:

Materials
Reinforcement: normally preimpregnated
  (SMC or DMC)
Resins: polyester, epoxy, vinyl ester
Fillers: up to 40%
Core materials: no

Tolerances
Overall dimensions: 0.2–1.0 mm
Thickness: 0.2–0.5 mm
Moulded detail: 0.1–0.2 mm

Part geometry
Thin skin parts with moulded-in ribs and
  bosses
Must allow mould separation 1° draw
  angle

Advantages
Very high production rate
Fine detail, close tolerance
Low cost
Long tool life

Process parameters
Fibre volume fraction: 0.12–0.4
                                    Size range: 0.1–2.5 m²
Processing pressure: 50–150 bar
Processing temperature: 130–150 °C
Production rate: 4–30 parts/hour

Cost factors
Production cost: Very low
Material cost: Low
Tooling cost: Very high
Equipment cost: Very high

Design details
Minimum radii: 0.1 mm
Ribs: yes
Bosses: yes
Holes: can be moulded in line of draw
Thickness changes: yes

Disadvantages
Mechanical properties modest with SMC
and DMC
Material flow causes property variability
High tooling cost

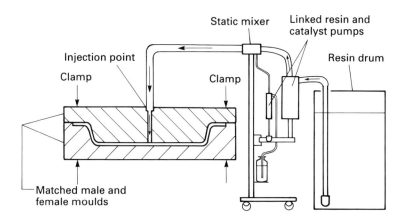

2.9 Resin injection.

Table 2.9 Composites process data sheet: resin injection

Composites process data sheet

Name: Resin injection | **Type:** Closed mould

Other names: resin transfer moulding

Abbreviations: RI
                   RTM

See also: cold press moulding
               vacuum assisted resin injection

Materials
Reinforcement: continuous strand mat or
    chopped strand preform
Resins: polyester, polyurethane, low
    viscosity epoxy
Fillers: not recommended
Core materials: Foams
                    Balsa

Tolerances
Overall dimensions: 1.0–2.0 mm
Thickness: 0.25–0.5 mm
Moulded detail: 0.25–0.5 mm

Part geometry
Complex shapes can be formed using multi-
part moulds, but best for thin-walled open
shapes

Advantages
Good surface both sides
Accurate dimensions
Wide range of part geometry
Reasonable production rate
Sandwich construction possible

Process parameters
Fibre volume fraction: 0.1–0.15
Size range: 0.25–5.0 m^2
Processing pressure: max. 2 bar
Processing temperature: 20–50 °C
Production rate: 1–4/hour/mould

Cost factors
Production cost: Moderate
Material cost: Moderate
Tooling cost: Moderate
Equipment cost: Moderate

Design details
Minimum radii: 1.0 mm recommended
Ribs: must be foam-filled
Bosses: not recommended
Holes: below 50 mm; cut or drill
          above 50 mm; cut or mould
Thickness changes: yes

Disadvantages
Massive tooling
Low fibre content

---

and low fibre content, but it is capable of making more complex
shapes than cold press moulding at a similar production rate.

A data sheet for resin injection is given in Table 2.9.

2.3.8 Vacuum-assisted resin injection

Vacuum-assisted resin injection is a development of resin injection
which overcomes many of the latter's limitations. The major benefits
are the ability to produce large mouldings and higher fibre content, and
the freedom to use high strength reinforcements. Better consolidation
and lower void content are also possible.

As with resin injection the reinforcement pack is loaded and the
mould closed. A seal is formed around the edge of the mould and a

Table 2.10 Composites process data sheet: vacuum-assisted resin injection

---

Composites process data sheet

Name: Vacuum-assisted resin injection	**Type:** Closed mould

Other names: none	**Process parameters**
	Fibre volume fraction: 0.15–0.35
Abbreviations: VARI	Size range: 1.0–30 m²
	Processing pressure: max. 2 bar
See also: resin injection	Processing temperature: 15–30 °C
vacuum bag	Production rate: 1–15/shift/mould
Materials	
Reinforcement: continuous strand mat	**Cost factors**
woven cloths	Production cost: Moderate
Resins: polyester, polyurethane	Material cost: Moderate
Fillers: not recommended	Tooling cost: Low
Core materials: Foams	Equipment cost: Low
Balsa	
Tolerances	**Design details**
Overall dimensions: 2.0–5.0 mm	Minimum radii: 3 mm
Thickness: 0.5–2.0 mm	Ribs: must be foam-filled
Moulded detail: 1.0–2.0 mm	Bosses: no
	Holes: cut or drill
Part geometry	Thickness changes: yes
Best suited to large thin-walled parts with	
no sudden changes in shape or thickness	
Advantages	**Disadvantages**
High fibre content	One-shot process; mods and repair difficult
Large size	Production development usually needed on
Low cost, low weight tooling	each mould
Full range of reinforcements	

---

partial vacuum introduced within the mould cavity. This consolidates the pack and removes air. Mixed resin is then injected by pump or gravity feed. A key feature of the process is the use of a flexible GRP upper mould which deforms under injection pressure to allow the resin to pass and is then reformed to its proper shape by vacuum when injection is complete. This ensures a high fibre content and good impregnation.

As vacuum is used to clamp the moulds, the construction of the moulds does not need to be as substantial as for basic resin injection.

A data sheet for vacuum assisted resin injection is given in Table 2.10.

2.3.9 Injection moulding

The injection moulding technique is shown diagrammatically in Fig. 2.10 and is the same capital-intensive, high volume process used to manufacture most thermoplastic components and many everyday products made in thermosetting plastics such as electric plugs and sockets.

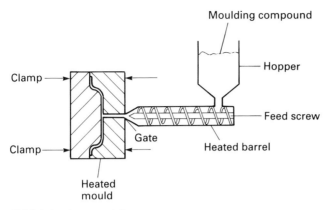

2.10 Injection moulding.

When used for making composite parts a dough must first be mixed containing all ingredients including the reinforcement. This is loaded into the moulding machine hopper and processed in essentially the same way as unreinforced materials. This means that the dough is forced into the mould by a screw or piston and this causes degradation of the fibres. Only short, random fibres can be used and their orientation

Table 2.11 Composites process data sheet: injection moulding

Composites process data sheet	
Name: Injection moulding	**Type:** Closed mould
Other names: none	**Process parameters** Fibre volume fraction: 0.05–0.1
Abbreviations: none	Size range: 0.01–1.0 m² Processing pressure: 750–1500 bar
See also:	Processing temperature: 140 °C Production rate: 10–60/hour/mould
Materials Reinforcement: random only	**Cost factors**
	Production cost: Very low
Resins: polyester compounds	Material cost: Low
Fillers: up to 30% by weight	Tooling cost: Very high
Core materials: no	Equipment cost: Very high
Tolerances Overall dimensions: 0.1–0.5 mm Thickness: 0.1 mm Moulded detail: 0.05 mm	**Design details** Minimum radii: 0.05 mm Ribs: yes Bosses: yes Holes: usually moulded
Part geometry Almost any solid shape	Thickness changes: yes
Advantages Very high production rate Very close tolerances Complex shapes	**Disadvantages** Very high tooling cost Mechanical properties limited and variable Size limit

in the mould cavity is determined by the flow during filling. Properties of parts made by injection moulding therefore tend to be variable.

A data sheet for injection moulding is given in Table 2.11.

2.4 Continuous processes

2.4.1 Continuous laminating

In the continuous laminating process, reinforcement and resin are combined and contained between two layers of release film which act as carriers transporting the laminate on a conveyer through a curing oven. On emerging from the oven the release film is peeled off and the cured laminate cut to length. Figure 2.11 shows the process.

The process can be used to make either a flat sheet or, by passing the material over formers prior to curing, a corrugated profile.

Resin impregnation is either by passing the reinforcement through a bath before the release films are attached or by applying a layer of resin to each release film via a 'doctor' blade and then sandwiching the

Table 2.12 Composites process data sheet: continuous laminating

Composites process data sheet		
Name: Continuous laminating	**Type:** Continuous	
Other names: none	**Process parameters**	
	Fibre volume fraction: 0.1–0.25	
Abbreviations: none	Size range: Up to 2.0 m wide	
	Processing pressure: low	
See also:	Processing temperature: 100–150 °C	
	Production rate: Up to 100 m/h	
Materials		
Reinforcement: random or woven	**Cost factors**	
Resins: polyester	Production cost:	Very low
Fillers: up to 20% by weight	Material cost:	Low
Core materials: no	Tooling cost:	Low
	Equipment cost:	High
Tolerances	**Design details**	
Overall dimensions: 1.0 mm	Minimum radii: 1.0 mm	
Thickness: 0.5–1.0 mm	Ribs: no	
Moulded detail: 0.5–1.0 mm	Bosses: no	
	Holes: drill or cut	
Part geometry	Thickness changes: no	
Flat or corrugated sheet		
Advantages	**Disadvantages**	
Very high production rate	Shape limitation	
Low cost		
Consistency		
Good mechanical properties		

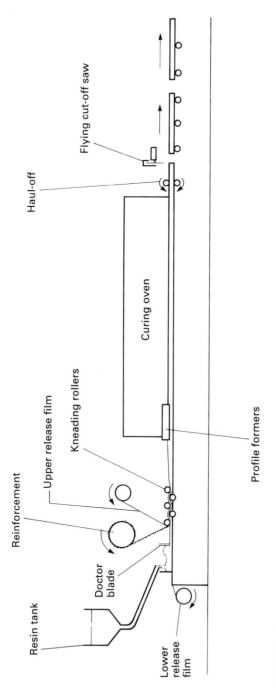

2.11 Continuous laminating.

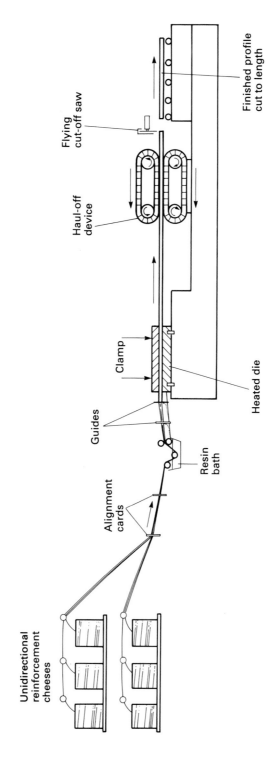

2.12 Pultrusion.

reinforcement. In either case the laminate and its carriers will subsequently pass through nip rollers to complete the consolidation and control the thickness.

This is a highly productive and capital-intensive method of manufacture but it is of course limited to flat sheet and simple profiles.

A data sheet for continuous laminating is given in Table 2.12.

2.4.2 Pultrusion

In the pultrusion process, shown in Fig. 2.12, reinforcement is impregnated with resin and pulled through a heated die which forms it to the final profile shape. It emerges from the die fully cured and ready to be cut to length. Resin impregnation can be by submerging the reinforcement in a bath before it enters the die or by injection directly into the die.

Pultrusion enables very high fibre contents to be achieved using continuous yarn, woven cloth or mat reinforcement. The tension applied to the fibres ensures good longitudinal alignment and hence very high and consistent longitudinal mechanical properties can be attained.

Table 2.13 Composites process data sheet: pultrusion

Composites process data sheet

Name: Pultrusion	**Type:** Continuous

Other names: none	**Process parameters**
	Fibre volume fraction: 0.3–0.65
Abbreviations: none	Size range: Up to 1000 mm wide
	Processing pressure: varies
See also:	Processing temperature: 130–150 °C
	Production rate: 10–30 m/h
Materials	
Reinforcement: all types	**Cost factors**
Resins: polyester, epoxy, vinyl ester	Production Cost: Low
Fillers: up to 25% by weight	Material cost: Moderate
Core materials: no	Tooling cost: High
	Equipment cost: High
Tolerances	**Design details**
Overall dimensions: 0.2–0.5 mm	Minimum radii: 0.1 mm
Thickness: 0.1–0.2 mm	Ribs: yes
Moulded detail: 0.1 mm	Bosses: no
	Holes: cut or drill
Part geometry	Thickness changes: yes
Almost any constant section	
Advantages	**Disadvantages**
High mechanical properties	Shape limitation
High production rate	Limited transverse properties
Close tolerance	
Consistency	

Open and closed profiles can be produced to close tolerances and quite intricate detail can be achieved. Multicellular profiles are also possible.

A data sheet for pultrusion is given in Table 2.13.

2.4.3 Continuous filament winding

Continuous filament winding is a method of producing filament wound pipe continuously. A winding head containing several 'cheeses' of continuous yarn reinforcement rotates around the mandrel, wrapping on yarn as it goes. The pipe emerges at a constant rate from a curing oven, the mandrel being of an ingenious design which continuously collapses onto itself and then reforms into a cylinder at the start of the process. More than one winding head can be used to create additional fibre angles.

A data sheet for continuous filament winding is given in Table 2.14.

Table 2.14 Composites process data sheet: continuous filament winding

Composites process data sheet

Name: Continuous filament winding	**Type:** Continuous

Other names: none

Process parameters
Fibre volume fraction: 0.55–0.7

Abbreviations: none

Size range: Up to 2.0 m diameter
Processing pressure: ambient

See also: filament winding

Processing temperature: ambient
Production rate: up to 2 m/min

Materials
Reinforcement: continuous rovings only
Resins: polyester, epoxy, vinyl ester
Fillers: can be used
Core materials: no

Cost factors
Production cost: Low
Material cost: Low
Tooling cost: High
Equipment cost: High

Tolerances
Overall dimensions: 1.0–2.0 mm
Thickness: 1.0–2.0 mm
Moulded detail: no

Design details
Minimum radii: not applicable
Ribs: no
Bosses: no
Holes: drill or cut
Thickness changes: no

Part geometry
Circular tubes only

Advantages
Excellent mechanical properties
High production rate
Consistency

Disadvantages
Circular tubes only
Spiral winding only

3   Finishing methods

3.1 Introduction

Mouldings produced by many of the processes described in Chapter 2 require to be trimmed, machined or coated in some way after moulding. The following sections discuss various techniques for trimming and finishing composite components.

3.2 Cutting

3.2.1 Cutting-trimming/deflashing

Most composite components emerge from the mould with excess material attached which needs to be removed in order to bring the part to its finished size. The thickness and composition of this 'flash' varies according to the moulding process and tooling accuracy.

For a precision hot press or injection moulded part the flash is usually a fraction of a millimetre thick and contains little or no reinforcement. It can normally be removed easily with a trimming knife, rasp or coarse abrasive paper.

3.2.2 Sawing

Where the flash is more than a millimetre thick, or contains a high proportion of reinforcement, a more robust cutting tool will be required. The most suitable hand tool is a standard hacksaw with a metal-cutting blade, and this can also be used for general cutting operations on most composites.

To speed up the rate of cutting some form of power tool is necessary. The jigsaw is a versatile and effective tool but relatively slow. To achieve a high rate of cut on thermoset composites, the high speed abrasive disc cutter should be used, and a diamond impregnated cutting wheel· is essential to achieve a reasonable wheel life.

Circular saws can be used to cut composites, but both saw and workpiece must be firmly held to avoid binding. This means that the saw needs to be bench-mounted and the workpiece held in a traversing table or vice versa. A diamond impregnated saw blade is necessary.

For more intricate cutting operations a high speed router or water jet cutter is recommended.

Most of the cutting methods discussed above involve high surface speed abrasion, and the inevitable result of this is the generation of large volumes of very fine dust. Whilst this is not carcinogenic, efficient dust extraction at source is essential to provide tolerable working conditions.

3.2.3 Routing

High speed routing is a very effective method of trimming and shaping thermoset composites. Surface cutting speeds of at least 125 metres per minute are recommended which means rotary speeds of at least 12 000 rpm, while 18 000 rpm is recommended. Typical feed rates will be in the range of 1–5 metres per minute. Diamond impregnated router bits are recommended although diamond pattern tungsten carbide tipped (TCT) bits can be used.

Hand routing is possible but imprecise; it is far better to use a fixed head or broken arm router or better still a CNC router. Standard woodworking machines are quite suitable providing care is taken to prevent abrasive or conductive dust entering the controls and bearings.

3.2.4 Drilling

Drilling composites with conventional high speed steel drill bits is straightforward, but achieving a clean hole is not. To prevent the drill biting into the surface, a reverse angle must be ground on the leading edge. Experiments with speed and feed rates will then usually be necessary to obtain best results. There are special drill bits available for cutting composites and for high fibre content materials they may be the only way to produce high quality holes.

Diamond-tipped hole saws are suitable for cutting large holes but material and saw should be held firmly. A better result will probably be achieved with a router.

3.2.5 Planing

The use of a hand or high speed rotary planer is not recommended with composites: blades tend to dig in and will be blunted very quickly. For hand shaping the 'Surform' type of plane or file can be used but blade life is again likely to be short.

The most practical way to produce a flat machined surface on a composite is with a belt or disc sander using a carborundum-faced cutting paper.

3.2.6 Turning

Wood or metalworking lathes can be used to turn composite parts but care must be taken to hold the part rigidly; this is best done where practical by leaving it on the mould for the turning operation.

It is difficult to achieve a high quality turned finish at normal cutting speeds as the material comes away unevenly. High speed turning or a rotating cutter will be necessary for the best finish.

3.3 Surface finishing

3.3.1 Buffing

Composite materials generally respond well to conventional polishing techniques, especially if there is a resin-rich surface layer. Coarse carborundum papers can be used dry if the surface is particularly rough, followed by wet or dry papers used with liberal quantities of soapy water to prevent clogging. The rubbing can be finished with 1000 grade paper, then the surface must be washed down with water and dried off, before being polished with automotive cutting compound such as T-Cut. A final wax polish will normally give a deep lustre.

Where there is a high fibre content in the surface layers avoid using coarse papers if possible as they will cut into the reinforcement and damage the material. In this case it is better to wet or dry lightly to remove projections, T-Cut lightly then work at building up a thick wax layer.

3.3.2 Etching

Etch primers are available; these are sold specifically to prepare composite materials for subsequent painting. They must be selected and used with extreme caution as it is difficult to limit chemical attack to the surface layers and there is thus a real danger of destroying the structural resin layers.

The best preparation for painting composites is to abrade the surface thoroughly as described below.

3.3.3 Painting

Despite the excellent surface finish and corrosion resistance of composites, it is often necessary to apply a surface coating as a post-moulding operation. A considerable amount of work has therefore been put into painting composite materials, particularly in the automotive industry. Special primers have been developed to allow composite parts to be top-coated alongside metal parts in conventional paint lines. In such high volume applications fully automatic painting is common.

When dealing with lower quantities the rules for preparation are:

1 Remove all traces of release agents, wax and grease by solvent cleaning.
2 Remove all surface gloss by abrading with fine wet or dry paper used with soapy water.
3 Wash with soapy water and rinse thoroughly, then dry out thoroughly.

Automotive cellulose and polymer lacquers can be used (with a suitable primer) but for a tough and long-lasting finish, a two-pack polyurethane applied directly with no primer works well.

3.4 Joining

3.4.1 Bonding

Adhesive bonding of composite components is a wide subject which can only be touched on in this chapter. Epoxy, polyurethane and acrylic adhesives are all used extensively and specialist advice should be sought on the type to be used in a particular application.

Satisfactory bonded joints depend on the correct choice of adhesive, proper preparation and careful application. Surfaces to be joined must be dry, clean and free from contaminants, particularly dust, oil and wax. If the base material is fully cured it may be necessary to abrade it to obtain a good key. Glue thickness is another critical factor in determining the choice of adhesive and the performance of the finished joint.

The design of bonded joints is covered in Chapter 8.

3.4.2 Riveting

Composites in general are brittle materials and are therefore not able to yield in the same way as metals in order to relieve local stress concentrations. For this reason the conventional expanding shank type of rivet is not recommended. However, such rivets are in common use with composites but result in considerable local damage to the laminates being joined and often work loose or pull through.

To avoid overstressing the laminate, a large contact area and a clenching action that pulls the laminates together without expanding the rivet shank are needed. A number of rivets of this type are available, some designed specially for use with laminated sheet materials such as composites.

Hole sizes must be large enough to allow some expansion of the rivet even if it is not of the expanding shank type.

Special care needs to be taken where laminate surfaces are uneven, such as in contact moulded parts: the rivet will need to have a wide range of grip lengths to accommodate thickness variations and 'quilting' may occur between rivets if non-moulded surfaces are in contact.

3.4.3 Bolting

As composite materials do not yield, care needs to be taken wherever stress concentrations occur, and this applies particularly to bolted connections. Load transfer through bolted joints is dealt with in some detail in Chapter 8: in practical terms it is important to match hole and bolt sizes accurately and to ensure a clean finish on bolt and hole surfaces if the full strength of the joint is to be realised. Also, because of the highly directional nature of the materials, the strength of joints is very dependent on the reinforcement type, content and direction relative to the applied load.

Bolted joints in composites can lose tightness over a period, consequently, some form of locknut or stiffnut is advisable, particularly if the connection is subject to cyclic or tensile loading.

Moulded-in holes are an economical way of achieving accurate size and positioning but the full strength of the joint is unlikely to be developed as it is extremely difficult to maintain the reinforcement alignment and content close to the hole.

Part III

Structural design of polymer composites

This part is divided into four chapters: the methodology and management of a design project, procedures for designing materials, the structural component design techniques and the limit state design method. The emphasis is on design methods for simple laminate analyses and component design and is based upon formulae associated with specially orthotropic laminates (i.e. those that are symmetric about their mid-plane with balanced reinforcement) and design charts and tables. Attention is directed towards mechanical behaviour of units; the in-service properties are not discussed. The mechanical properties of the composites are assumed to be modified to take account of anisotropy and viscoelastic behaviour, such as creep, and of the rate of dependence of polymer materials. A consequence of this is that the polymer material data values are reduced from that of their short term values.

The design of composites is an interactive process between the material design, the structural design and the manufacturing techniques for the composite material/structure. These three processes must be considered simultaneously because of the many different fibre arrays that can be incorporated into the matrix material and the diverse manufacturing methods that are available, all of which will affect the final product in terms of ultimate failure under certain loading conditions.

Only stress analyses of the materials and structures under load have been discussed, but it should be remembered, however, that the criteria for design of composite beams are generally based upon displacement, although complete collapse of the structural system is stress-related.

Factors of safety have been incorporated into the limit state design discussions in Chapter 7.

4 Methodology and management of a design project

4.1 Introduction

Competent designers, engineers and analysts can all execute their trade if given the data and design methodologies. That is what this book is all about: providing the data and design methodologies for composite materials.

There are, however, a number of design considerations particular to these materials of which those new to the subject need to be aware. There is also the need for a great deal more interaction between the members of the design team, because of the many influencing factors. This puts a greater emphasis on the role of design management to control all the activities in order to achieve an optimum solution.

It is also essential that a quantified design process is followed in order to avoid overlapping of skills and to ensure that the right skills are introduced at the most appropriate point in the design programme.

The designers need therefore to have not only a good technical grounding in the subject, but also a sound knowledge of all the other activities, an understanding of how the design process works and a sympathy with the needs of good management.

4.2 Design considerations

A polymeric composite material is made up of at least two materials: a fibre and a matrix. These are combined to exploit the individual characteristics, thereby providing additional qualities that they are unable to provide individually.

They differ markedly from metals in the following ways:

1 Composites are mostly orthotropic and inhomogeneous.
2 Generally, stiffness is less than that of steels, leading to greater attention to local and overall structural stability.
3 Material properties are influenced by the manufacturing process, temperature and the environment.
4 Most resins are combustible.
5 Material costs, particularly for the high strength and stiffness fibres, form a high proportion of the product cost.

Furthermore, when comparing composite materials to metals it is found that:

1 They are lighter, leading to excellent specific strength and stiffness values.
2 They have very good environmental resistance and do not corrode like many metals.
3 Fibres can be used strategically leading to ease in optimising weight.
4 They are readily formed into complex shapes.
5 They have low thermal conductivity.

With the above points in mind, the designer new to the subject of composite materials needs to address a number of particular areas.

4.2.1 Designing the laminate

Many structural materials generally have isotropic properties and they are homogeneous, that is to say, they are uniform in all directions.

A composite material can take a number of different forms. The material may be orthotropic, such as a unidirectionally reinforced polymer, where the strength and stiffness in the fibre direction considerably exceeds that at 90° to the fibre. It may be planar-isotropic, such as a random chopped strand glass mat reinforced polymer. It may approach isotropy by the use of very short fibres randomly placed in a polymer by injection moulding. In all cases, though, composite materials are inhomogeneous.

It is these anisotropic properties of composite materials that are the key to developing highly efficient structures. Fibres can be strategically placed so that they locally engineer the required strength and stiffness properties. Furthermore, by combining different fibre types – glass, aramid, carbon, etc – the particular properties of each fibre can be exploited. For instance, the low cost of glass, the extreme toughness of aramid fibre and the high strength and stiffness of carbon can all be used within a single laminate.

A composite material is not ductile like metal, and failure, when it occurs, is abrupt. The stiffness properties are generally lower than those of steel, but the lower weight of composite materials results in excellent specific strength and stiffness properties, leading to reduced-weight components and structures.

The properties of the laminate are affected by the amount of fibre in the matrix, which in turn is influenced by the manufacturing process.

4.2.2 Establishing property data

Obtaining reliable data for any material is essential to the success of the design function. Obtaining reliable data for composite materials is aggravated by fragmentation of suppliers. One set of companies produces the fibre, while others produce the matrix or resins. Further companies produce yet more materials – prepregs, sandwich cores, bonding agents, etc. It is therefore quite often difficult to obtain reliable data for a combination of materials that makes up the finished laminate. Furthermore, the properties will be affected by the manufacturing process and the working environment.

It generally therefore falls to the designers to create their own data bank of properties by testing materials on an ongoing basis. This is an essential function within the design process and a useful data bank can be slowly achieved.

Reliance on manufacturer's data, where available, requires an element of caution. Often the data are not adequately quantified and testing standards are not made available.

4.2.3 Designing for the environment

The environmental factors of heat, cold, moisture, ultraviolet light and aggressive materials (such as acids) can all have an adverse effect on the performance of a composite laminate over a period of time. The extent of the effect is a function of the fibres and resins selected, and, of course, the degree of the environmental condition.

Needless to say, composite materials behave extremely well in many arduous environmental situations. The marine and chemical industries have shown this to be the case where fibre reinforced polymers are accepted materials. But care is needed, particularly in a hot, wet atmosphere where the mechanical performance of some polymeric composites can fall by as much as half from those at normal ambient condition. This requires particular attention if long term loading conditions are envisaged.

Adequate testing of materials in the expected environment is the only way to ensure a satisfactory structural integrity.

4.2.4 Designing for joints and assemblies

Composites have the principal advantage of being chemically joined during their manufacture. Consequently, a carefully manufactured large component need not have any joints requiring mechanical fasteners or secondary bonded joints. However, it is not always possible, desirable or even practical to make a single unit and more often than not a complete assembly is obtained from a number of components or sub-assemblies.

To this end, the designer needs to have a good understanding of the performance of mechanical joints, which are often governed by the strength of the matrix as opposed to the fastener itself. Consequently, long term loads may be affected by the creep properties of the matrix.

Alternatively, use can be made of bonded joints, which has the advantage of less concentration of load through the joint. Provided care is exercised in avoiding stress concentrations caused by abrupt changes in section, excellent joints can be designed that are suitable for bonding, and made in a properly controlled environment.

4.2.5 Designing for robustness and through life performance

Robustness is difficult to quantify as it is a measure of the ability of a structure or component to survive knocks, shocks and rough handling. Composite materials can be shown to have good resilience and toughness, particularly aramid reinforced polymers which are also favoured in ballistic missile protection. The ability of a structure to withstand through-life impacts can be enhanced by good manufacturing quality, eliminating production flaws such as voids, excessive air inclusions and shrinkage cracks, but good design plays an equally important role. The correct selection of materials, the provision of adequate load paths and the avoidance of stress concentrations are the most important aspects.

Composite materials generally have good fatigue resistance and they can be used to overcome through-life problems often associated with metals, particularly where the metal has been welded.

4.2.6 Designing for manufacture

A most important aspect of designing in composite materials is the interaction with the manufacturing processes, which are varied and diverse. The designer needs to have a good understanding of how the many processes work and how the design will be affected by the process. For instance, the quality control achievable with contact moulding falls very short of that available with an oven-cured fibre and resin system. Consequently, factors of safety used in design need to reflect these differences and the selection of design stresses may need to reflect the variation in properties that will occur.

It is also essential that the design process incorporates a very early involvement from those people who are truly versed in the manufacturing processes, as an incorrect choice could have disastrous results with respect to the product's performance and cost.

4.2.7 Designing for cost

Expensive materials do not necessarily lead to expensive components. Reduced handling costs reduce product cost, reduced weight enhances the through-life performance, increased environmental resistance prolongs life, etc.

Composite materials have many unique characteristics which the designer can exploit to advantage, but cost must be carefully quantified. For any dynamic structure or component, this must include a through-life cost assessment to ensure that the best use is made of any reduced weight.

4.3 The need for design management

Design management is now a recognised and respected activity but has only recently received the attention it deserves. The driving factor has been the need to produce better and more cost effective products.

The simple objective of design management is to ensure that the product meets the set parameters and is fit for the purpose intended within an accepted cost envelope. With the introduction of composite materials as the manufacturing material, this objective is brought sharply into focus because of the wide range of candidate constituent materials – fibres, resins, sandwich cores, etc. – the range of cost of these materials, the range of process techniques, assembly and quality control procedures – all of which can detract and often confuse the design team. Design management ensures that a logical and controlled interactive procedure is followed, allowing the various disciplines to achieve the best compromise for the sake of an efficient and cost effective product.

The function and necessity for design must be understood at all levels, from the corporate to the individual.

At the corporate level, the overall planning is undertaken by senior management, defining what is needed and balancing this with costs and available resources. This corporate level must also understand the importance of the design function – it is not an overhead, it is a company asset.

At the management level, project managers ensure that the particular task is properly controlled and managed. The project objectives are defined and the necessary disciplines are brought together.

At the design level, the designers use all available resources, data and facilities to design the product, within an accepted and controlled design methodology.

All this may sound fairly obvious, but the above design considerations highlight the many critical influencing factors which make management of the design so important when composite materials are involved. It is also important to understand and use a recognised design process.

4.4 The design process

It is necessary to have a framework in which design can function. The introduction of composite materials as the primary material enhances this necessity. An attempt to rationalise the design process for composite

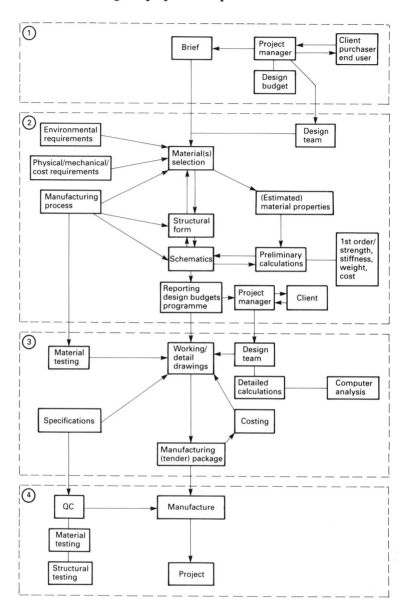

4.1 Design process for composite structures (after Ref. [4.1]).

materials has been previously made [4.1] and this is graphically illustrated in Fig. 4.1.

The process is divided into four phases. The first is perhaps the most important: the brief. Without a well-defined brief, no product can be adequately designed. In fact, it should be argued that without a brief, no product should be commenced. The effect of the brief on composite materials is important, particularly with respect to cost. The cost of composites can range by a factor of 100. The brief therefore must contain comprehensive information on the cost envelope, including the through-life or operational costs. The required quantity is also important because this in turn will affect the cost by way of tooling, cycle times, materials used, etc.

The cost of design is also an aspect of the first phase which requires addressing carefully. As previously stated, design should not be treated as an overhead, but as an asset. Consequently, it should be costed and included in the overall costs. A design team working to a properly costed programme is one that is better working to an open budget. The associated discipline is not only good for the company but it also helps those in the design team value their own time.

The second phase involves the preliminary design. It is in this phase that the importance of design management must be appreciated. There are many disciplines involved, several activities and, with composites, many different candidate materials and processes. It is extremely easy to go off in several directions when looking for an acceptable design solution, especially if those involved do not take the advice of their fellow professional disciplines.

All designers should ask themselves at the early stages of a design programme: can it be made and can it be assembled? With composites, these questions need more attention because of the many processes available, each with its own advantages and disadvantages. For instance, closed mould technology (RTM, etc.) provides a moulded surface to each face, as opposed to contact moulding or female tooling and prepregs. But closed mould technology has certain restrictions on geometry, such as draw angles, which, unless properly addressed, may lead to problems in manufacture and assembly.

Also of importance in this second phase is the question of property data. One aspect of composites not always initially appreciated by designers used to metals, is that the material is often made at the point of manufacture, as opposed to buying in materials and fabricating. The material's strength and stiffness are therefore functions of the process, the quality control during the process operation, the skill of the operators and, of course, the constituent materials used. The data used in the preliminary design therefore have to be assumed or estimated theoretically. Experienced designers will have created their own data banks of material properties from previous projects and they will be able to make a selection of design properties with confidence. Less experienced designers must use published data, and care is needed in using some data, particularly those which do not reference the source or which do not fully quantify the properties, e.g. the absence of the volume fraction.

When the design concept and the manufacturing process have been frozen, the ongoing design costs have been checked and the brief reassessed, the third phase is entered: that of detail design. During this phase, the selected materials need to be used with the specified manufacturing process to make materials for testing to establish the mechanical properties, thereby reaffirming the estimated properties from the second phase, or to revise the design for the actual values.

The interaction of skills can perhaps be appreciated when the second and third phases of the design process are studied. The composite industry is a fluid industry. It is being introduced to new materials, processes and end uses constantly. No one person can therefore be sufficiently equipped to cope with all these factors at once. The interaction of skills is therefore essential in order to bring the best of all these factors into the design process.

This should not, and does not, stop when the final stage is entered; that of manufacture. More so than with other materials, the design process for composite materials must continue throughout the manufacturing period. The performance of the product is affected by the process, the quality control, inspection, product maintenance and handling. Feedback from this phase is important information to put back into the design envelope.

4.5 Design potential

Composites offer the designer unique possibilities to improve the performance of a product leading to cost effective engineeering. They have excellent specific strength and stiffness properties, and through-life environmental resistance, they can be formed into complex shapes and they have the ability to be tailored to suit local and overall mechanical performance parameters.

There are, of course, some disadvantages, such as the high cost of the high performance materials and the combustibility of many of the resin systems, but perhaps the biggest disadvantage is the lack of design standards, methodologies and quality control procedures, and the lack of understanding by both management and designers of the materials themselves.

All these aspects are now being addressed and it is certain that the industry will grow as more and more applications are established, which will in turn provide the impetus for improved standards and understanding.

The designer is at the heart of this progression and the possibilities for composites are not only considerable, but also exciting. Progression must, however, be undertaken within a controlled framework and the importance of the management of design and the need for an established design process with accepted design methodologies requires constant emphasis.

4.6 References

[4.1] A.F. Johnson and A. Marchant, *Design and Analysis of Composite Structures*, Chapter 3, 'Polymers and polymer composites in construction'. Edited by L. Hollaway, 1990, Thomas Telford, London.

5 Procedures for designing materials

5.1 The principles of design analysis

As in all structural designs the developed stress and strain levels in the polymer composite when it is under load must be determined and the material designed. The critical stress, strain and deformation values are then compared with the relevant design criteria to ensure that the component satisfies product requirements and material limitations. Polymer composites are usually macroscopically inhomogeneous and anisotropic because of the reinforcing fibres and, in addition, have viscoelastic properties derived from the polymer matrix. Owing to the differing material descriptions between composites, further material properties are required to characterise polymer composites completely, consequently, more complex analysis procedures are required to determine stress and deformation levels than are generally required for the more conventional materials.

The three main aspects of material design which will be considered are:

1 The analysis which considers the anisotropy and inhomogeneity in polymer composites (the material properties of the fibre and matrix, ply orientation, layer thicknesses, etc.).
2 The short term load condition, in which the elastic stress and analysis methods may be used, provided anisotropy is taken into account.
3 The long term load conditions, in which viscoelastic and degradation effects may be significant; in this case it would be necessary to modify the short term elastic design procedures.

5.2 Requirements of materials' design

Polymer composite materials generally consist of laminae of resin impregnated fibres which are unidirectionally or orthogonally aligned, angle-ply or randomly orientated systems. It is also possible to provide a mixture of fibre arrays in adjacent laminae when fabricating a composite material to meet the required loading situation. This freedom to tailor-make composite materials with specific required properties introduces an additional complexity in the design analyses of these systems over those of the conventional ones.

As the design of composite structures ideally involves the simultaneous analysis and design of the material and the structural system, this approach may be undertaken by the finite element analysis. It can be expensive for small jobs and is really relevant only to the high technology of the aerospace industry; for the medium technology applications a simpler approach is to consider the material design independently from that of the structural one. Consequently, for the latter design application, the properties of a chosen fibre/matrix array are calculated or measured and are then utilised in the structural analyses.

The majority of polymer composite structural systems are composed of relatively thin plates or shell laminates where the properties may be in terms of laminate structure and ply thickness using laminated plate theory or by commercially available PC software.[5.1–5.3] Assuming that the laminates had orthotropic symmetry and that both in-plane direct and shear loads as well as bending and twisting moments were acting on the plate, see Fig. 5.1, the element properties would require two principal tensile stiffnesses, shear stiffness and two principal flexural rigidities. In addition, the corresponding strength values in tension, flexure and shear would be required; the latter three values would be obtained by either mechanical tests or by undertaking a laminate analysis and thus the laminate stiffness and strength characteristics would be known. To satisfy the necessary design criteria this relatively small number of properties would then be used in the structural analysis and design for the composite.

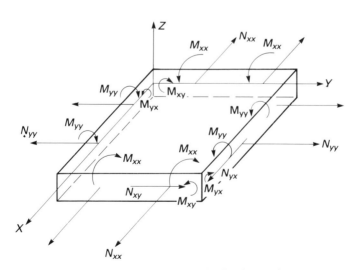

5.1 Schematic diagram of a composite laminate element.

5.2.1 Short term loading considerations

A typical tensile stress–strain characteristic of a randomly orientated glass fibre–polymer composite, shown in Fig. 5.2, would indicate an approximate linear elastic relationship when a relatively low stress level is applied to the composite. This will eventually progress to a non-linear situation which indicates the onset of micro-damage (fibre debonding, resin cracking or transverse ply failure) and brittle failure; the latter is associated with fracture of the fibres.

Polymer composites may be characterised as:

1 Linear-elastic for design purposes, and the relevant elastic moduli and failure stresses would then be incorporated into the calculations. If necessary, reduced values of stress would be used to take account of micro-damage failure.

2 Orientated fibre composites in which the properties are highly dependent upon the direction of fibres; the design would then consider the anisotropic nature of the material.

The composite material design will be divided into two main groups, those that may be considered quasi-isotropic and those that are anisotropic.

The *randomly orientated fibre polymer* material is the most widely used quasi-isotropic composite in engineering applications and the mechanical properties which are required for design are the modulus of elasticity, the Poisson ratio and a failure stress. The modulus of rigidity would be obtained from the relation between E and v. Thus:

$$G = E/2(1+v) \tag{5.1}$$

The *anisotropic polymer composite* materials would be those with specific fibre orientations as unidirectionally aligned fibre plies or woven

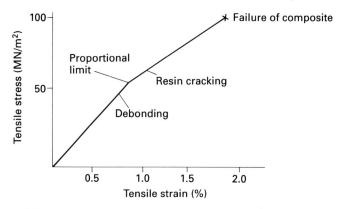

5.2 Tensile stress–strain curve for typical randomly orientated fibre/matrix laminate under short term loads.

fibre fabric. It is highly likely that those composites would possess orthotropic material properties by virtue of the fact that the fibres, and hence the composite, would possess two orthogonal symmetry directions in the plane of the laminate. Four elastic constants would therefore be required to characterise the material stiffness properties. These are the moduli of elasticity of the composite in directions 1 and 2 (E_{11} and E_{22}), the modulus of rigidity (G_{12}) and the Poisson ratio v_{12}. The usual notation has been adopted here, see Fig. 5.3, where the first suffix refers to the plane upon which the stress acts and the second suffix is

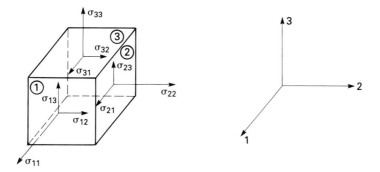

5.3 General stress system.

Table 5.1 Design data for typical polymer composite materials

Properties	Randomly orientated glass fibre–polyester	Woven roving glass fibre–polyester	Unidirection-ally aligned glass fibre–polyester	Unidirection-ally aligned aramid fibre–polyester	Unidirection-ally aligned carbon fibre–epoxy
v_f	0.2	0.35	0.6	0.7	0.7
E_{11} (GPa)	8	16	42	76	180
E_{22} (GPa)	8	16	12	8	10
G_{12} (GPa)	3	4	5	3	7
v_{12}	0.3	0.15	0.3	0.34	0.28
σ_{11}* ult (MPa)	95	250	700	1400	1500
σ_{11}* dam† (MPa)	50	100	—	—	—
σ_{22}* ult (MPa)	95	250	30	12	40
σ_{22}* dam (MPa)	50	100	—	—	—
σ_{12}*	80	95	72	34	68

* ultimate failure
† damage stress

the coordinate direction in which the stress acts. As laminates are two-dimensional systems, only stresses in the planes 1,2 are considered. The convention for the Poisson ratio v_{12} is that the second suffix refers to the strain produced in direction 2 when the lamina is loaded in direction 1. The Poisson ratio v_{21} is derived from the equation.

$$v_{21} = v_{12} \, E_{22}/E_{11} \hspace{4cm} [5.2]$$

The ultimate strengths σ^*_{11}, σ^*_{22} and σ^*_{12} are also required to characterise the material. Typical tensile strengths and moduli of elasticity values are given in Table 5.1 for randomly orientated woven rovings and unidirectionally aligned Kevlar and carbon fibre composites. It is important to remember that these values are only guides as the actual values will depend not only upon the type of fibre used but also upon the fibre volume fraction of each type of reinforcement and the technique used for the composite manufacture. The corresponding values for composites under compressive and flexural loads will be different from those given above owing to the inhomogeneity in polymer composites. It should be noted that the above values are those for ultimate failure; however, matrix cracking stresses may be considered the failure criteria, in which case a lower failure stress would be quoted.

The orthotropic properties of, say, woven rovings in the off-axes are much lower than those in the principal fibre directions and in design the lower stresses, particularly the shear stresses, need to be addressed.

Failure in the unidirectionally aligned composites could occur in the transverse direction as a result of low secondary stresses transverse to the main fibre direction and if this is a particular problem composites with cross-ply or angle-ply reinforcement are utilised.

5.2.2 Long term loading considerations

Polymer composites will not behave linearly under long term loading because of the viscoelastic time, frequency and rate dependent properties of the polymer material; the creep component will be particularly significant in materials with low fibre volume fractions. Under dynamic fatigue load, complex interactions take place between fibres and matrix leading to a degradation in composite mechanical properties. Both of these long term aspects of polymer composites need to be addressed by the designer.

There are particular situations where an elastic analysis may be used as the basis for design calculations. If this procedure is adopted, elastic design results are used but these may be modified to take account of the complex polymer behaviour and thus make use of the extensive design data and formulae for elastic materials derived for small strain theory. This approach would be acceptable for filled and fibre reinforced polymers which have only slight viscoelastic behaviour under short term loads. If, however, viscoelastic effects are significant, a pseudo-elastic design method may be applied to these materials where the elastic modulus and other elastic constants are replaced by a time or rate dependent modulus. Williams[5.4] has used the elastic design method and Ogorkiewicz[5.5] has discussed the pseudo-elastic method applied to the creep calculations.

The basis upon which the pseudo-elastic method is formed is to approximate the viscoelastic stress–strain response to a set of pseudo-elastic stress–strain curves in which the modulus of elasticity and failure stress properties, for instance, are no longer constants but may be time, rate or frequency dependent. The pseudo-elastic material properties are then used in the normal elasticity-based design procedures to predict the behaviour of the component at a particular time, rate or frequency. This enables orthotropic elastic design formulae or finite element analysis to be undertaken for design of long term loading situations provided that the relevant and measured material property data are used.

A set of long term moduli (viz. E_{11}, E_{22}, G_{12} v_{12}) are required for a design analysis of a polymer composite unit under a long term load; these will normally be obtained from creep tests on the material. The measured stain–time function $\varepsilon(t)$ may be used to define the time dependent modulus [$E_{11}(t) = \sigma/\varepsilon(t)$]. The creep rupture stress $\sigma*_{11}$ will be obtained from the creep rupture–time curve. See Section 2.9 of ref. [5.6].

During a dynamic fatigue loading of polymer composites, resin cracking and fibre–matrix debonding could be initiated. This microdamage could eventually lead to ultimate failure of the composite at stresses below the short term strength. In a fatigue test the results of the number of cycles to fail a specimen at different stress loads are presented on an S–N curve which shows the relationship between the failure stresses and the number of load cycles under a range of load conditions. Although there is some reduction of the modulus of the composite during the fatigue loading, generally it is not considered significant to require a modification of the stiffness characteristics of the material in the design analysis. Table 5.2 shows typical data for the magnitude of long term load effects on three types of composites; the results are presented as percentages of short term values and demonstrate the importance of allowing for long term effects in design analyses.

Knowing the long term strength and stiffness values of the composite, as a percentage of the short term, the pseudo-elastic method can be used for design analyses in the conventional short term design procedures to analyse the composite under load for time t or after a fixed load cycle N. This procedure can be formulated as partial material coefficients; this is illustrated in the limit state design method (Chapter 7). A similar approach is used in the BS 4994[5.7] for composite tanks, by using the long term property reduction factors as design coefficients for creep and cyclic loading; analogous factors are also used to allow for the degradation due to environmental exposure.

Table 5.2 Typical long term load reduction factors for GFRP: data refer to the ratio of long term strength/short term strength at ambient test conditions (from ref. [5.9]).

Long term strength reduction factor	Randomly orientated glass fibre-polyester	Woven roving polyester	Unidirectionally aligned glass fibre epoxy
Fatigue stress ratio			
10^3 cycles	0.6–0.7	0.35–0.45	0.4–0.6
10^6 cycles	0.25–0.35	0.20–0.30	0.2–0.3
Creep rupture stress ratio (10^2 hours)	0.65–0.75	0.55–0.65	0.75–0.90

5.3 Characteristics of Materials

Many polymer composites are manufactured by stacking a series of lamina which are separated by resin-rich areas in a predetermined arrangement to ensure optimum properties and performance. This system is shown in Fig. 5.4. There are, of course, load variations in material structure which have a small effect upon the mechanical properties but when these are being determined for design analysis it is assumed that each lamina, and consequently the composite, is macroscopically homogeneous and therefore has uniform elastic properties and volume fraction throughout. It follows that the composite material properties are a smeared value of all lamina. If there is no basic ply data it would be necessary to determine composite properties in terms of those of the fibre and matrix. This could be undertaken by utilising theoretical formulae ranging from the simple rule of mixture to more rigorous analyses of the complete set of elastic constants as discussed

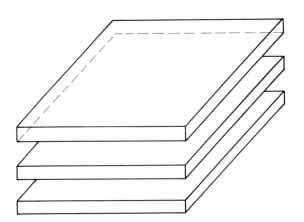

5.4 Composites manufactured by stacking laminates.

in ref. [5.7]. In addition, material property values may be estimated by feeding into microcomputer software packages[5.1-5.3] basic values of fibre strength, stiffness, volume fraction and ply angle.

In this discussion only specially orthotropic laminates (those which are symmetric about their mid-plane with balanced reinforcement) will be considered. These systems are commonly used in practice and are simpler to analyse compared with the general orthotropic lamina. Four in-plane stiffnesses and four flexural stiffnesses are required to define the properties of a specially orthotropic laminate under the general load system shown in Fig. 5.1.

1 Consider the laminate behaviour with respect to the x–y plate element coordinate axes (see Fig. 5.5) where θ is equal to zero such that these axes lie in the material symmetry axes 1–2.

The in-plane direct and shear loads N_{xx}, N_{yy} and N_{xy} are shown in Fig. 5.1 and the corresponding in-plane strains $\varepsilon_{xx}, \varepsilon_{yy}, \varepsilon_{xy}$ for a specially orthotropic laminate are given by:

$$\begin{bmatrix} N_{xx} \\ N_{yy} \\ N_{xy} \end{bmatrix} = \begin{bmatrix} A_{11} & A_{12} & 0 \\ A_{21} & A_{22} & 0 \\ 0 & 0 & A_{66} \end{bmatrix} \begin{bmatrix} \varepsilon_{xx} \\ \varepsilon_{yy} \\ \varepsilon_{xy} \end{bmatrix} \qquad [5.3a]$$

i.e. $[N] = [A][\varepsilon]$ where the four in-plane stiffnesses have values of:

$$\begin{aligned} A_{11} &= E_{11}h/\{1-(v_{12}{}^2E_{22}/E_{11})\} \\ A_{22} &= E_{22}h/\{1-(v_{12}{}^2E_{22}/E_{11})\} \qquad [5.3b] \\ A_{12} &= v_{12}E_{22}h/\{1-(v_{12}{}^2E_{22}/E_{11})\} \\ A_{66} &= G_{12}h \end{aligned}$$

The bending and twisting moments, M_{xx}, M_{yy}, M_{xy} also shown in Fig. 5.1 with their corresponding curvatures $\chi_{xx}, \chi_{yy}, \chi_{xy}$ for specially orthotropic laminates, are given by:

$$\begin{bmatrix} M_{xx} \\ M_{yy} \\ M_{xy} \end{bmatrix} = \begin{bmatrix} D_{11} & D_{12} & 0 \\ D_{21} & D_{22} & 0 \\ 0 & 0 & D_{66} \end{bmatrix} \begin{bmatrix} \chi_{xx} \\ \chi_{yy} \\ \chi_{xy} \end{bmatrix} \qquad [5.4a]$$

i.e. $[M] = [D][\chi]$

where four laminate flexural rigidities have values of:

$$\begin{aligned} D_{11} &= E_{11}h^3/12(1-v_{12}{}^2E_{22}/E_{11}) \\ D_{22} &= E_{22}h^3/12(1-v_{12}{}^2E_{22}/E_{11}) \qquad [5.4b] \\ D_{12} &= v_{12}E_{22}h^3/12(1-v_{12}{}^2E_{22}/E_{11}) \\ D_{66} &= G_{12}h^3/12 \end{aligned}$$

It will be seen from the above that the in-plane stiffnesses and the flexural rigidities are defined in terms of the laminate structure and ply properties.

For a generally orthotropic or anisotropic laminate the equivalent

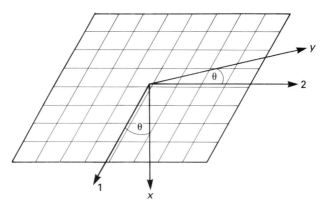

5.5 The material axes 1–2 and the plate axes x–y.

results of equations [5.4a] and [5.4b] can be written succinctly as follows:

$$\begin{bmatrix} N \\ M \end{bmatrix} \quad \begin{bmatrix} A & B \\ B & D \end{bmatrix} \quad \begin{bmatrix} \varepsilon \\ \chi \end{bmatrix}$$ [5.4c]

where the symmetry about the leading diagonal can be noted. The [A] matrix is the extensional stiffness matrix relating the in-plane stress resultants [N] to the mid-surface strains [ε] and the [D] matrix is the flexural stiffness matrix relating the stress moments to the curvatures [χ]; these are the equations [5.4a] and [5.4b] respectively. As the [B] matrix relates [M] to [ε] and [N] to [χ] it is called the bending-extensional coupling matrix. It should be noted that a laminate structure can have bending-extensional coupling even if all the lamina are isotropic. The only time that the structure is exactly symmetric about its middle surface is when all the [B] components are equal to zero and this requires symmetry in laminate properties, orientation and location from the middle surface. If coupling did take place between the in-plane and bending properties, other properties may be required. These are discussed in refs. [5.7] and [5.8].

Having obtained the values of laminate in-plane loads and moments from equations [5.3] and [5.4] respectively, the corresponding failure loads and bending moments under uniaxial conditions are given in terms of the relevant ultimate stress values as:

$$\begin{aligned}
N^*_{xx} &= h\sigma^*_{11} & M^*_{xx} &= h^2\sigma^*_{11}/6 \\
N^*_{yy} &= h\sigma^*_{22} & M^*_{yy} &= h^2\sigma^*_{22}/6 \\
N^*_{xy} &= h\sigma^*_{12} & M^*_{xy} &= h^2\sigma^*_{12}/6
\end{aligned}$$ [5.5]

2   If the material principal directions (the 1- and 2-directions) do not coincide with those of the co-ordinate axis, in which the application of

the load lies, transformation equations must be used to relate stresses from axes 1–2 to those in the axes x–y. The derived formulae for the transformed components of $[A]$ and $[D]$ stiffness matrix in items of $E_{11}, E_{22}, G_{12}, v_{12}, h$ and θ (the angle between the axes 1 and x, see Fig. 5.4) can be found in ref. [5.7] and [5.8].

In design analysis, the modulus of elasticity and tensile strength of a laminate under a uniaxial load applied in the x-direction at an angle θ to the axis's 1-direction may be determined from the above analysis of the transformed stiffness matrix. The modulus E_{xx}, E_{yy}, G_{xy} is given in ref. [5.7] as:

$$\frac{1}{E_{xx}} = \frac{m^4}{E_{11}} + \frac{n^4}{E_{22}} + \left[\frac{1}{G_{12}} - \frac{2v_{12}}{E_{11}}\right] n^2m^2$$

$$\frac{1}{E_{yy}} = \frac{n^2}{E_{11}} + \frac{m^4}{E_{22}} + \left[\frac{1}{G_{12}} - \frac{2v_{12}}{E_{11}}\right] n^2m^2 \qquad [5.6]$$

$$\frac{1}{G_{xy}} = 2\left\{\frac{2}{E_{11}} + \frac{2}{E_{22}} + \frac{4v_{12}}{E_{11}} - \frac{1}{G_{12}}\right\}n^2m^2 + \frac{1}{G_{12}}\{n^4 + m^4\}$$

where $m = \cos\theta$ and $n = \sin\theta$.

The significance of the material anisotropy in polymer composites can be seen in the relationships of E or σ as a function of the orientation angle θ, shown in Fig. 5.6(a) and (b) respectively for a unidirectional fibre–polymer composite ($\simeq$ 75% fibre by weight) woven roving fibre–polymer matrix composite ($\simeq$ 50% fibre by weight) and a randomly orientated fibre–polymer matrix composite ($\simeq$ 25% by weight).

The application of a uniaxial load in the x-direction of an orthotropic material will cause tensile and shear stress components in the 1–2 axes (the fibre axis) and the tensile failure σ^*_{xx} will be dependent upon the multiaxial failure criteria adopted for the material. However, currently there is no universal failure criterion valid for polymer composite materials and the two theories for failure most widely used are the maximum stress and the quadratic stress criteria.

The maximum stress theory of failure assumes that failure occurs when the stress in the principal material axis reaches a maximum value. There are three possible modes of failure and the conditions for these are:

$$\sigma_{11} = \sigma^*_{11}$$
$$\sigma_{22} = \sigma^*_{22} \qquad [5.7]$$
$$\sigma_{12} = \sigma^*_{12}$$

where σ^*_{11} and σ^*_{22} are the ultimate tensile or compressive stress in direction 1 or 2 respectively and σ^*_{12} is the ultimate shear stress acting in plane 1 in direction 2.

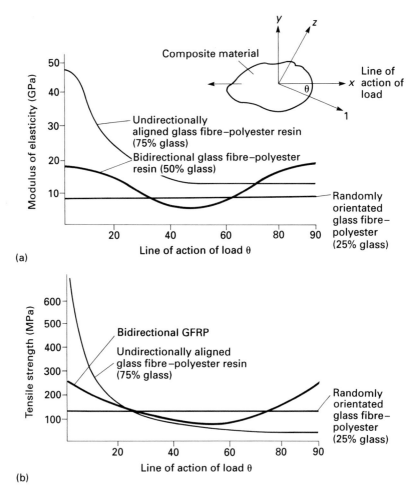

(a)

(b)

5.6 Typical relationship of (a) modulus of elasticity and (b) tensile strength and line of action of load for different types of reinforcement.

If the load were applied to the lamina at an angle θ to the principal axis's direction, then by transformation,

$$\sigma_{11} = \sigma_{xx}\cos^2\theta = \sigma_\theta\cos^2\theta$$
$$\sigma_{22} = \sigma_{xx}\sin^2\theta = \sigma_\theta\sin^2\theta \qquad [5.8]$$
$$\sigma_{12} = -\sigma_{xx}\sin\theta \cos\theta = -\sigma_\theta\sin\theta\cos\theta$$

The off-axis ultimate tensile strength σ_θ is the smallest value of the following stresses:

$$\sigma_u = \sigma^*_{11}/\cos^2\theta \text{ or } \sigma^*_{22}/\sin^2\theta \text{ or } \sigma^*_{12}/\sin\theta\cos\theta \qquad [5.9]$$

One form of the quadratic stress criterion (the Tsai–Hill criterion) is based upon the strain energy in the composite, and failure occurs when the following equation is satisfied.

$$\frac{(\sigma_{11})^2}{(\sigma^*_{11})^2} - \frac{\sigma_{11}\sigma_{22}}{(\sigma^*_{11})^2} + \frac{(\sigma_{22})^2}{(\sigma^*_{22})^2} + \frac{(\sigma_{12})^2}{(\sigma^*_{12})^2} = 1 \qquad [5.10]$$

This equation assumes that the composite is transversely isotropic. In most composite materials $\sigma^*_{11} \gg \sigma^*_{22}$ so that the second term in the above equation tends to zero and the equation becomes:

$$\frac{(\sigma_{11})^2}{(\sigma^*_{11})^2} + \frac{(\sigma_{22})^2}{(\sigma^*_{22})^2} + \frac{(\sigma_{12})^2}{(\sigma^*_{12})^2} = 1 \qquad [5.11]$$

Using the Tsai–Hill equation [5.10] in conjunction with equation [5.9] a prediction can be made of the failure strength in direction θ to the principal axis of an unidirectional laminae by the use of the design equation:

$$\sigma_\theta = \left\{ \frac{\cos^4\theta}{(\sigma^*_{11})^2} + \left[\frac{1}{(\sigma^*_{12})^2} - \frac{1}{(\sigma^*_{11})^2} \right] \sin^2\theta \cos^2\theta + \frac{\sin^4\theta}{(\sigma^*_{22})^2} \right\}^{-1/2} \qquad [5.12]$$

5.4 General Laminates

Polymer composites tend to be manufactured by laminating plies of polymer-fibre laminates together, as shown in Fig. 5.4, and the requirements of the composite will dictate the laminate arrangement. The continuous unidirectionally aligned and woven roving plies which carry the main load would normally be positioned symmetrically about the centre of thickness of the composite and the randomly oriented fibre-polyester with a lower fibre-matrix ratio would be positioned at the surface so that the polyester would provide a greater corrosion barrier to protect the load-bearing laminates. The randomly oriented fibre-polymer composites are also used between the woven roving composites to improve the interlaminar adhesion and strength.

In laminated polymer composites it is necessary to compute the stiffness properties [A], [D] and the failure loads in terms of the composite structure, ply orientations and the component material properties.

There are several commercial software packages available (see refs [5.1–5.3] for the design of composites and they are all based upon plate theory. The programs are available in several disk versions for running on microcomputers (e.g. Macintosh, IBM PC and compatibles). The input data to the program are the material properties of the components of the composite and failure stresses. In addition, ply thickness, ply orientation and geometric values are required. The output data from the programs will contain values of elastic moduli and flexural rigidities, ply stresses and strains and any possible failure. It is also possible to compute first ply failure when the laminate is damaged and last ply failure when the laminate has failed. Analyses of various influences on the laminate can be undertaken, including the effect of temperature on cooling during manufacture and moisture uptake which causes swelling.

The designer, where possible, should use the commercial software packages for the analysis and design of laminates because of the complexity of the mechanical behaviour of the material. However, for widely used laminates it might be possible to employ simplified design formulae but such laminates would tend to be unidirectionally aligned fibre reinforced composites with cross-ply or angle-ply reinforcement.

5.4.1 Cross-ply laminates

Cross-ply laminates are fabricated with unidirectional fibres being placed at 0° and 90° to the Cartesian co-ordinate axes. If in addition the cross-ply laminate is symmetric about the mid-plane thickness it will have specially orthotropic properties. Figure 5.7(a) shows the arrangement of a cross-ply laminate and the relative positions of the axes (1,2) and (x,y).

Using the laminate plate theory:

$$A_{11} = [h/(1-v_{12}{}^2E_{22}/E_{11})][VE_{11}+(1-V)E_{22}]$$
$$A_{22} = [h/(1-v_{12}{}^2E_{22}/E_{11})][(1-V)E_{11}+VE_{22}] \qquad [5.13]$$
$$A_{12} = v_{12}hE_{22}/(1-v_{12}{}^2E_{22}/E_{11})$$
$$A_{33} = hG_{12}$$

where h = thickness of laminate and V = volume fraction of 0° plies.

For the commonly used balanced laminate composed of n plies of equal thickness (h/n) and where the volume fraction (v) in the 0° and 90° are equal, the flexural rigidities can be calculated from the following equation (from ref. [5.8]):

$$D_{11} = \{h^3/[12(1-v_{12}{}^2E_{22}/E_{11})]\}\{\tfrac{1}{2}[E_{11}+E_{22}]+(3/2n)(E_{11}-E_{22})\}$$
$$D_{22} = \{h^3/[12(1-v_{12}{}^2E_{22}/E_{11})]\}\{\tfrac{1}{2}[E_{11}+E_{22}]-(3/2n)(E_{11}-E_{22})\}$$
$$D_{12} = v_{12}E_{22}h^3/[12(1-v_{12}{}^2E_{22}/E_{11})] \qquad [5.14]$$
$$D_{33} = h^3G_{12}/12$$

It will be evident that the flexural properties in equation [5.14] are dependent upon the number of plies; for a few plies the values of D_{11} and D_{22} are significantly different. However, as the number of plies increases, the value of the term $(3/2n)$ becomes small, the value D_{11} tends to that for D_{22} and the cross-ply material becomes as a homogeneous material with modulus $1/2 [E_{11} + E_{22}]$ in both flexure and tension.

Therefore, equation [5.14] for a large number of plies, where the volume fraction V is 0.5, becomes:

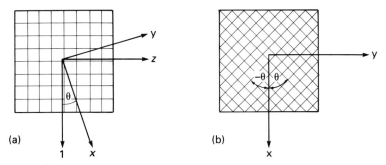

5.7 Arrangement of (a) a cross-ply and (b) an angle-ply laminate.

$$D_{11} = D_{22}$$
$$= \{h^3/[12(1-\nu_{12}{}^2E_{22}/E_{11})]\}\{\tfrac{1}{2}(E_{11}+E_{22})\} \tag{5.15}$$

For the general orthotropic laminate the problem is much more complicated and there are no simple design formulae for accurately predicting laminate strengths. For the simplest loading condition where the laminate is under a uniaxial stress situation, the plies will generally be subjected to a multiaxial stress field. This will require a check on the stress condition of each ply as well as the introduction of a multiaxial failure criterion. As discussed earlier, this investigation is best undertaken by the utilisation of a laminate software package, although for preliminary analysis, Johnson,[5.9] has introduced some simplifying assumptions regarding the laminate stress fields and modes of failure.

Consider a cross-ply laminate under axial tensile load in the Cartesian co-ordinate direction x in which the zero degree plies lie (see Fig. 5.7(a)).

The initial damage will usually consist of transverse ply cracking and this will occur when σ_{xx} reaches the value of the transverse ply strength σ^*_{22}. Therefore investigating the case when the laminate is balanced, symmetric and composed of a large number of plies of equal volume placed in the 0° and 90° directions (i.e. $V = 0.5$). Then the loads of transverse ply failure are:

$$N^*_{xx} = N^*_{yy} = h\sigma^*_{22}[1/2E_{22}][E_{11}+E_{22}] \tag{5.16}$$

If the load is increased beyond this value, ultimate failure of the laminate will occur when the longitudinal plies fail and if it is assumed that no further load is taken by the cracked transverse plies then the load to failure is carried entirely by the longitudinal fibres. This is the condition assumed for the rule of mixtures values. The ultimate tensile failure of the cross-ply laminate is:

$$N^*_{xx} = N^*_{yy} = 1/2h\sigma^*_{11} \tag{5.17}$$

Considering this laminate to be under a flexural loading and that damage will take place in the surface transverse plies or the one adjacent to it, the bending moment for the balanced symmetric cross-ply laminate at first ply failure is:

$$M^*_{xx} = M^*_{yy} = h^2\sigma^*_{22}\{1/12E_{22})(E_{11}+E_{22})\} \qquad [5.18]$$

Because it is not possible to know the state of the transverse plies (some may still be carrying load) at the ultimate flexural load, no flexural equation can be developed. It is likely, however, that the flexural strength will be higher than that of the tensile strength. In detailed analysis it would be necessary to use the computer-based laminate analyses.

5.4.2 Angle-ply laminates

Figure 5.7(b) illustrates an angle-ply laminate in which the various plies are oriented at $\pm\theta$ to the plate element axes, in this case the x-axis. These laminates, which have an equal member of $+\theta$ and $-\theta$ plies are balanced about their mid-plane, are orthotropic in nature. In the case where they have a large number of thin plies arranged as above they are specially orthotropic. In this case, the in-plane and bending stiffnesses are given[5.7] as:

$$(A_{11}, A_{12}, A_{22}, A_{33}) = h(\bar{Q}_{11}, \bar{Q}_{12}, \bar{Q}_{22}, \bar{Q}_{33}) \qquad [5.19]$$
$$(D_{11}, D_{12}, D_{22}, D_{33}) = (h/12)(\bar{Q}_{11}, \bar{Q}_{12}, \bar{Q}_{22}, \bar{Q}_{33}) \qquad [5.20]$$

where the functions A and $\bar{Q}$ are given as:

$$\bar{Q}_{11}=1/(1-v_{12}^2E_{22}/E_{11})\{E_{11}m^4+[2E_{22}v_{12}+4G_{12}(1-v_{12}^2E_{22}/E_{11})]m^2n^2+E_{22}n^4\}$$
$$\bar{Q}_{22}=1/(1-v_{12}^2E_{22}/E_{11})\{E_{22}m^4+[2E_{22}v_{12}+4G_{12}(1-v_{12}^2E_{22}/E_{11})]m^2n^2+E_{11}n^4\}$$
$$\bar{Q}_{12}=1/(1-v_{12}^2E_{22}/E_{11})\{E_{22}v_{12}(n^4+m^4)+[E_{11}+E_{22}-4G_{12}(1-v_{12}^2E_{22}/E_{11})]n^2m^2$$
$$\bar{Q}1/(1-v_{12}^*E_{22}/E_{11})\{[G_{12}(1-v_{12}^2E_{22}/E_{11})](n^4+m^4) + E_{11}+E_{22}-2G_{12}(1-v_{12}^2$$
$$E_{22}/E_{11})m^2n^2\} \qquad [5.21]$$

where m and n are $\cos\theta$ and $\sin\theta$ respectively and θ is the ply angle.

The failure of angle-ply laminates is complex, with the failure modes being dependent upon the angle of the ply.

If the laminate is loaded uniaxially in direction x (i.e. the plate axis, see Fig. 5.7(b)) the initial damage is likely to be shear cracking in the plies, particularly if these plies have $\theta < 45°$. The ply shear stress σ_{12} in the principal axes 1–2 for a load in the x-axis direction is obtained from the transformation relationship as:

$$\sigma_{12} = \sigma_{xx} \sin\theta \cos\theta \qquad [5.22]$$

Therefore, at first ply failure, because of shear cracking, an estimation of the in-plane load N^*_{xx} and an out-of-plane bending moment M^*_{xx} is

$$N^*_{xx} = h\sigma^*_{12} \sin\theta \cos\theta \text{ and } M^*_{xx} = \frac{1}{6} h^2\sigma^*_{12} \sin\theta \cos\theta \quad [5.23]$$

where σ^*_{12} is the unidirectionally aligned ply shear strength.

The ultimate load will occur at fibre fracture and this can be estimated by netting analyses (see ref. [5.10]) in which the load in the composite is assumed to be carried by the fibres. This leads to the following estimates of the ultimate loads in an angle-ply laminate, thus:

$$N^*_{xx} = h\sigma^*_{11}\cos^2\theta, \quad M^*_{xx} = \frac{1}{6} h^2\sigma^*_{11}\cos^2\theta$$

These formulae are acceptable for small ply angles where the netting assumptions are valid.

5.5 Sandwich construction materials

Sandwich construction materials are used if it is necessary to increase the stiffness of the overall cross-section of the composite or the individual laminates whilst still maintaining a low weight panel. A sandwich construction material normally consists of fibre reinforced polyester laminates bonded to a low density core material; Fig. 5.8 illustrates this form of construction. The face materials (i.e. the reinforced polyester laminates) support the flexural loads and axial forces within the composite cross-section and the core material supports the majority of the shear. The core material may be low density polymer and the four types of foam used in civil engineering (although these are not exhaustive) are: (a) rigid polyurethane, (b) phenolic, (c) polyvinyl chloride and (d) polystyrene.

The first two are thermosets and the last two are thermoplastic materials. In the high technology industries, such as the aerospace, carbon fibre–epoxy polymer composites in conjunction with aluminium honeycomb or resin-impregnated paper honeycomb are employed. In addition integral skin polyurethane foam is used in automotive panel applications. These sandwich materials allow expensive composites to

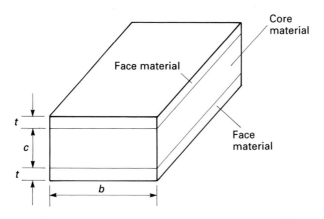

5.8 Sandwich construction.

be utilised simultaneously and effectively as structural and as thermal insulation units.

It is clear then that the structural sandwich beams and panels are efficient systems for flexural loading. The stiff face materials are at the plate surface where flexural stresses are high and the low stress region may consist of a lower modulus and low density material to reduce the weight of the overall panel. Sandwich panels may be classified under two main headings:

1   The low shear modulus core sandwich.
2   The shear modulus core, which is sufficiently stiff to have only limited influence on panel deflections.

5.5.1 Sandwich beams associated with low shear modulus cores

The first type of sandwich beam has been discussed in length in refs [5.7] and [5.11]. Section 1.5.3.4 has discussed the properties of the thin face sandwich and, consequently, this section will summarise the main design formulae and general principles of the sandwich beam and panel. Type (1) can be divided into three groups:

1   A very thin face sandwich, where the ratio $d/t > 100$.
2   A thin face sandwich, where the ratio $100 > d/t > 5.77$ is satisfied.
3   A thick face sandwich, where the ratio $d/t < 5.77$ is satisfied.

The notation has been given in Fig. 5.8.

The designer of engineering structures is not generally interested in the first group and therefore this group of sandwich systems will not be discussed. The bending stiffness for the last two groups are:

$$D_f = E_f\,(btd^2)/2 \qquad\qquad\qquad [5.24]$$
$$D_f = E_f\,(btd^2)/2 + E_c(bt^3)/6 \qquad\qquad [5.25]$$

respectively, where E_f = modulus of elasticity of the face material and E_c = modulus of elasticity of the core material.

These formulae hold provided $(E_f/E_c)(t/c)(d/c)^2 > 16.7$ and $d/t > 5.77$ are fulfilled for the first relationship (equation [5.24]), and that the condition

$$(E_f/E_c)(t/c)(d/c)^2 > 16.7$$

is fulfilled for the second relationship (equation [5.25]).

The shear stiffness of a sandwich beam is:

$$N = AG = Gbd^2/c \qquad\qquad\qquad [5.26]$$

where $A = bd^2/c$ and G = shear modulus of core, and if $d/c \simeq 1$ the equation may be written as

$$AG = Gbd \qquad\qquad\qquad [5.27]$$

The shear stress in the core of the sandwich beams for the last two types are:

$$\tau = Q/bd \qquad\qquad [5.28]$$
$$\tau = Q/D \,[(E_f \, td)/2] \qquad\qquad [5.29]$$

respectively where Q = the shear force on the section under consideration.

The total deflections for sandwich beams is the sum of the deflections due to the bending moment and the shear deformations. This later deflection must be considered as it can be a significant percentage of the whole deflection; the two components which involve the bending and shear stiffnesses of the beam have been given above.

In considering a sandwich strut it is assumed that if the strut is narrow it will bend anticlastically whereas if it is wide it will bend cylindrically. An approximate correction may be made by replacing the material stiffness E_f by: $E_f/(1-v_f^2)$ where v_f = Poisson ratio of face material, and the bending stiffnesses are then:

$$D_1 = (E_f btd^2/2) \qquad \text{for the narrow strut [5.30]}$$
$$D_2 = (E_f btd^2/2(1-v_f^2)) \qquad \text{for the wide strut [5.31]}$$

In any analysis of the buckling of a sandwich strut the buckling stress is taken as the lower value of either (a) or (b) below where (a) is associated with the buckling stresses and (b) is associated with the wrinkling stresses.

(a) The critical load of a pin-ended strut with thin faces is:

$$P = P_E[1 + P_E/AG] \qquad\qquad [5.32]$$

where $P_E = \pi^2 D_1/L^2$, AG is given in equation [5.26] and D_1 is given in equation [5.30].

The critical load of a pin-ended strut with thick faces is:

$$P = P_E[1+(P_{Ef}/P_c)-(P_{Ef}/P_c)(P_{Ef}/P_E)]/[1+(P_E/P_c)-P_{Ef}/P_c] \qquad [5.33]$$

where $P_E = \pi^2 D_1/L^2$, $P_{Ef} = \pi^2 E_f I_f/L^2$, $P_c = AG$ (given in equation [5.26]).

(b) Wrinkling instability in sandwich struts (i.e. the compression faces of sandwich beams) occurs because of the instability associated with the short wave-length ripples on the faces. Because this is a buckling effect, the stresses may have smaller values than those predicted in formulae for struts under compression. Figure 5.9 shows the major types of wrinkling instability.

The stress in the sandwich faces at which wrinkling occurs is:

$$\sigma = B_1 \, E_f^{1/3} \, E_c^{2/3}$$

where B_1 = non-dimensionless buckling coefficient (see ref. [5.11]).

If the face material of the sandwich beam has initial irregularities, then under load these irregularities are increased and tensile stresses are developed in areas between the face and the core at these positions.

5.9 Principal types of wrinkling instability.

The ultimate tensile failure occurs when the bond between the faces and the core materials fails and this happens when the tensile stress reaches the value:

$$\sigma = B_2 \, E_f^{1/3} \, E_f^{2/3}$$

where B_2 = non-dimensional coefficient (see ref. [5.11]).

5.5.2 Sandwich beams associated with stiffer shear modulus cores

The materials used to obtain a greater shear stiffness sandwich beam compared with those discussed in Section 5.5.1 are:

1 A higher density polyurethane (PU) foam compared with that used in Section 5.5.1.
2 A higher density polyvinyl chloride (PVC) foam compared to that used in Section 5.5.1.

Section 6.4.1 discusses the greater stiffness sandwich beam and incorporates the material properties of the component parts into the design of the beam, consequently, for the design of materials reference should be made to Section 6.4.1 and subsequent sections.

5.6 Concluding remarks

This chapter has discussed how the material design properties of a composite element could be determined by calculation or by laboratory tests. The following chapter considers how these values for design formulation are used in composite structural design. When numerical analysis, such as finite element analysis (FEA), is used to determine the composite sizes and configurations, the material design property values are required as input data for the computer software. If orthotropic elastic material properties for the composite laminate or for the sandwich material are required, these will be obtained from analysis of the composite as discussed in Section 5.3. If composites consist of a single type of reinforcement, as discussed in Section 5.4, the material property data required for the numerical analysis will be the modulus of elasticity and the ultimate strength values as shown in Table 5.1. For the general laminates discussed in Section 5.4, the FEA would require homogeneous orthotropic properties and these would be obtained from the formulae given in Section 5.4.1 and 5.4.2 or by using a laminate software program as a preprocessor to the main FEA program.

5.7 References

[5.1] ESDU 2022 *Stiffnesses and Properties of Laminated Plates*, (ESDU Pac 8335); ESDU 2033 *Failure of Composite Laminates*, (ESDU Pac 8418), London, ESDU International, 1992.

[5.2] MIC-MAC Think Composites Software, Dayton, 1987.

[5.3] CoALA, Cranfield Institute of Technology 1990; WEBBER, Cranfield Institute of Technology 1990; ALLAN, Cranfield Institute of Technology 1990; LAMHOLE, Cranfield Institute of Technology 1990.

[5.4] J.G. Williams, *Stress Analysis of Polymers*, 2nd edition, Harlow, Longman, 1980.

[5.5] R.M. Ogorkiewicz, *Thermoplastics: Properties and Design*, Chichester, Wiley, 1974.

[5.6] L. Hollaway, *Polymer Composites for Civil and Structural Engineering*, Glasgow, Blackie Academic and Professional, 1993.

[5.7] BS 4994, *Design and Construction of Vessels and Tanks in Reinforced Plastics*, BS 4994, London, British Standards Institution, 1981.

[5.8] S.W. Tsai and H.T. Hahn, *Introduction to Composite Materials*, Westpoint, Technonic, 1980.

[5.9] A.F. Johnson, *Engineering Design Properties of GRP*, London, British Plastics Federation, 1984.

[5.10] D. Hull, *An Introduction to Composite Materials*, Cambridge, Cambridge University Press, 1981 (reprinted 1990).

[5.11] H.G. Allen, *Analysis and Design of Structural Sandwich Panels*, Oxford, Pergamon, 1969.

6 Structural component design techniques

6.1 Introduction

6.1.1 Preliminary design analysis

Design with composites is an iterative process (Fig. 6.1) which consists first of the selection of fibre and matrix material and laminate structure. Laminate properties obtained from measurement or from classical laminate theory (CLT) are then used as the materials data in a structural analysis, where the influence of structure geometry and loads are calculated, and structural failure criteria applied. The iterative step with composites is to change the laminate construction and to repeat the structural analysis in order to improve structural performance. In this way FRP materials are tailored for an application through the choice of fibre, matrix, fibre orientations, ply thickness, etc. Whilst this design freedom is an important factor in the success of FRP materials, it is the main reason for the increased complexity in design analysis with these materials.

Materials selection and laminate analysis were described in Chapters 1 and 5, and the main objective of this chapter is to describe structural design methods for composites. It is assumed here that laminate properties have been determined from a previous laminate analysis and it is shown how the results of a structural analysis give information about the influence of fibre orientation and laminate lay-up, and may be used to optimise the laminate construction. This is achieved by paying special attention to a *preliminary design analysis* based on simplified geometry and loads, where the influence of laminate construction, component geometry, load conditions and failure modes can be identified. This analysis is then used for the selection of the laminate and component geometry in the iterative design process. For more detailed design this preliminary design analysis may be followed by a finite element analysis (FEA) where geometrical details, fastenings, load transfer, etc., can be studied. For many less critical applications the simplified methods described here are sufficient.

In Sections 6.2–6.5 design procedures are given for composite struc-

Materials design

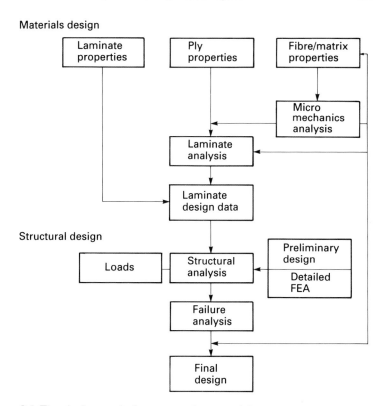

6.1 The design cycle for composite materials.

tural elements such as plates, thin-walled beams, sandwich plates and pressure vessels. They are based on simplified design formulae, design charts and tables which have been derived from a more detailed mathematical analysis of anisotropic structures. Examples are given to show how these procedures are used for FRP materials selection and for the design of composite structures. These procedures are given in a set of design data sheets written in note form with the minimum of explanatory text. Each data sheet has a diagram of the element geometry and loads, it lists the main design formulae, shows relevant design charts and concludes with some design tips and important references. This chapter is concluded in Section 6.6 with a brief discussion of detailed design of composite structures based on design software and FEA.

The literature on composites design concentrates on laminate analysis, however refs [6.1–6.3] discuss structural design with composites, [6.4] is a valuable source of general design formulae and [6.5] includes both CLT and the structural analysis of composite elements.

6.1.2 Composite structural element properties

Chapter 5 described how most composite structures are composed of relatively thin plate or shell laminates so that a thin laminated plate element (Fig. 6.2) can be taken as the basic structural element. When referred to the (x,y) axes in the plane of the element, the most general loading consists of in-plane direct and shear loads N_{xx}, N_{yy}, N_{xy}, bending moments M_{xx}, M_{yy}, and a twisting moment M_{xy}, all measured per unit width of element. Under these loads the element will stretch and bend giving rise to strains ε_{xx}, ε_{yy}, ε_{xy}, and curvatures χ_{xx}, χ_{yy}, χ_{xy}. It is assumed in this chapter that the structural element is (specially) orthotropic with material symmetry axes (x,y). The general laminate load–strain and moment–curvature relations, equation [6.4c] then decouple and take the forms:

$$\begin{bmatrix} N_{xx} \\ N_{yy} \\ N_{xy} \end{bmatrix} = \begin{bmatrix} A_{11} & A_{12} & 0 \\ A_{12} & A_{22} & 0 \\ 0 & 0 & A_{66} \end{bmatrix} \begin{bmatrix} \varepsilon_{xx} \\ \varepsilon_{yy} \\ \varepsilon_{xy} \end{bmatrix} \qquad [6.1]$$

$$\begin{bmatrix} M_{xx} \\ M_{yy} \\ M_{xy} \end{bmatrix} = \begin{bmatrix} D_{11} & D_{12} & 0 \\ D_{12} & D_{22} & 0 \\ 0 & 0 & D_{66} \end{bmatrix} \begin{bmatrix} \chi_{xx} \\ \chi_{yy} \\ \chi_{xy} \end{bmatrix} \qquad [6.2]$$

The laminate properties required for structural analysis are the in-plane stiffnesses $[A]$ and flexural rigidities $[D]$. Specially orthotropic laminates are characterised by four in-plane stiffnesses, A_{11}, A_{22}, A_{12}, A_{66}, and four flexural rigidities, D_{11}, D_{22}, D_{12} and D_{66}. Many commonly used composite structures such as cross-ply and angle-ply laminates, unidirectional (UD) composites, fibre fabric structures which are *balanced* and *symmetric*, and composed of a reasonable number of thin plies, are specially orthotropic with stiffness matrices having the forms indicated in equations [6.1] and [6.2]. The zero elements in $[A]$ and $[D]$ ensure that the laminate has no tension–shear and bending–torsion coupling. The coupling matrix in equation [5.4a] $[B] = 0$, hence there is no tension-bending coupling.

As an alternative to equation [6.1], the laminate in-plane strains may be characterised in terms of the averaged laminate stresses σ_{xx}, σ_{yy}, σ_{xy} and the effective engineering elastic constants for the laminate, by the equations

$$\begin{bmatrix} \varepsilon_{xx} \\ \varepsilon_{yy} \\ \varepsilon_{xy} \end{bmatrix} = \begin{bmatrix} \dfrac{1}{E_{xx}} & \dfrac{-v_{yx}}{E_{yy}} & 0 \\ \dfrac{-v_{xy}}{E_{xx}} & \dfrac{1}{E_{yy}} & 0 \\ 0 & 0 & \dfrac{1}{G_{xy}} \end{bmatrix} \begin{bmatrix} \sigma_{xx} \\ \sigma_{yy} \\ \sigma_{xy} \end{bmatrix} \qquad [6.3]$$

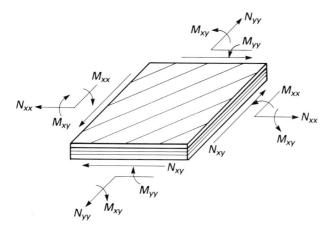

6.2 Schematic diagram of a composite laminate element.

Here E_{xx}, E_{yy} are laminate tensile moduli in the x- and y-directions, G_{xy} the in-plane shear modulus and υ_{xy} the lateral contraction ratio. In the convention used here, υ_{xy} is the strain in the y-direction per unit strain in the x-direction, for a load applied in the x-direction.

The following definitions and shorthand notation for the elastic constants and flexural rigidities are used extensively in this chapter and should be noted here:

$$v_{yx} = v_{xy}E_{yy}/E_{xx}$$
$$\mu = 1 - v_{xy}\,v_{yx} = 1 - v_{xy}^2\,E_{yy}/E_{xx} \qquad [6.4]$$
$$D_1 = D_{11}$$
$$D_2 = D_{22}$$
$$D_3 = D_{12} + 2D_{66}$$

and for a homogeneous orthotropic plate of thickness h:

$$D_1 = E_{xx}h^3/12\mu$$
$$D_2 = E_{yy}h^3/12\mu$$
$$D_{66} = G_{xy}h^3/12$$

Laminate strength properties under uniaxial loads are characterised by the element in-plane tensile and shear failure loads N_{xx}^*, N_{yy}^* and N_{xy}^*, and by the bending moments at failure M_{xx}^*, M_{yy}^* and M_{xy}^*, all measured per unit width of laminate. Alternatively under in-plane loads the averaged laminate failure stresses σ_{xx}^*, σ_{yy}^*, σ_{xy}^* may be required. These uniaxial strength values are the simplest to use in structural analysis calculations. In practice applied loads are seldom uniaxial and laminate strength values under a multiaxial load are needed.

As input data for the composite's structural analysis, it is assumed that a laminate lay-up has been selected and the laminate stiffnesses or effective elastic constants, together with uniaxial strength values, have been calculated from CLT or measured directly for the chosen laminate. The structural analysis calculates deflections and failure loads in the composite's structure, which may be used iteratively to modify the initial laminate lay-up. Other possible refinements include the use of the structural analysis to determine the multiaxial load conditions on a laminate element (e.g. Sections 6.2.2 and 6.5.2) so that an appropriate multiaxial laminate strength reserve factor can be calculated from CLT.

6.2 Design of composite panels

6.2.1 In-plane and buckling load

6.2.1.1 Uniaxial tension or compression
See Fig. 6.3.

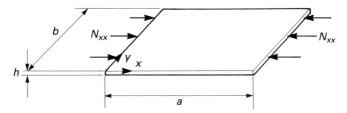

6.3 Panel under uniaxial compression load.

Geometry
Flat rectangular panel, sides a, b, thickness h.

Applied loads
Uniaxial tensile/compression load N_{xx} per unit plate width, $N_{yy} = N_{xy} = 0$.

Deformations
Plate strains are given in terms of the plate stiffnesses $[A]$ by equation [3.3.1]:

$$\varepsilon_{xx} = N_{xx}/(A_{11}-A_{12}^2/A_{22})$$
$$\varepsilon_{yy} = -\varepsilon_{xx}A_{12}/A_{22}$$
$$\varepsilon_{xy} = 0 \qquad\qquad [6.5]$$

and in terms of the plate engineering constants by equation [6.3] with $\sigma_{xx} = N_{xx}/h$:

$$e_{xx} = N_{xx}/hE_{xx}$$
$$\varepsilon_{yy} = -v_{xy}\varepsilon_{xx}$$
$$\varepsilon_{xy} = 0 \qquad\qquad [6.6]$$

Plate displacements follow directly from the strains, e.g. in the x-direction $u_x = a\varepsilon_{xx}$.

Failure loads

Tension: the tensile failure load is:

$$N_{xx}{}^* = h\sigma_{xx}{}^* \qquad [6.7]$$

where $\sigma_{xx}{}^*$ is the average laminate uniaxial failure stress as measured or as calculated from CLT with an appropriate failure criterion.

Compression: under uniaxial compression load the plate will usually buckle at a load N_{xx}', below the compressive strength load $N_{xx}{}^*$ in equation [6.7]. Buckling loads depend on plate geometry, edge support conditions and plate flexural properties.

For small aspect ratios $0.5 < a/b < 2$ and for simply supported edges, the uniaxial buckling load is:

$$N_{xx}' = (\pi^2/b^2)\,(D_1 b^2/a^2 + D_2 a^2/b^2 + 2D_3) \qquad [6.8]$$

For long, simply supported plates with $a/b > 2$, local buckling takes place which is independent of length a at a load:

$$N_{xx}' = 4\pi^2 D_1 K_1/b^2$$
$$K_1 = [(D_2/D_1)^{1/2} + D_3/D_1]/2 \qquad [6.9]$$

The plate failure load in compression is now:

$$\min (N_{xx}{}^*, N_{xx}').$$

Plates with clamped edges will have buckling loads much higher than equations [6.8] and [6.9], which are thus conservative values for design.

Shear
See Fig. 6.4.

Geometry
Flat rectangular panel, sides a, b, thickness h.

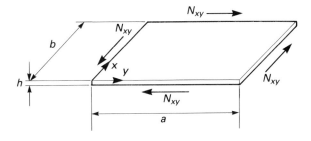

6.4 Panel under in-plane shear load.

Applied loads
In-plane shear load N_{xy} per unit plate width, $N_{xx} = N_{yy} = 0$.

Deformations
Plate strains are given directly from equation [6.1] or equation [6.3]:

$$\varepsilon_{xy} = N_{xy}/A_{66} = N_{xy}/hG_{xy}$$
$$\varepsilon_{xx} = \varepsilon_{yy} = 0 \qquad\qquad [6.10]$$

Failure loads
For shear material failures, the failure load is:

$$N_{xy}{}^* = h\sigma_{xy}{}^* \qquad\qquad [6.11]$$

where $\sigma_{xy}{}^*$ is the average laminate shear strength as measured or calculated from CLT. Thin composite plates will usually buckle in shear before this shear failure load is reached. Buckling loads depend on plate geometry, edge support conditions and plate flexural properties. For long plates with $a/b > 2$ and simply supported edges, local buckling occurs, which is independent of length a, at a shear load N_{xy}'.

$$
\begin{aligned}
N_{xy}' &= & 52K_2D_1/b^2 \\
K_2 &= & (D_2D_3)^{1/2}f(c)/D_1 \\
c &= & (D_1D_2)^{1/2}/D_3 \qquad\qquad [6.12] \\
f(c) &= & \left\{ \begin{array}{l} c^{1/2}(0.62 + 0.38/c), c \geqslant 1 \\ 0.89 + 0.04c + 0.07c^2, c < 1 \end{array} \right.
\end{aligned}
$$

The plate failure load in shear is now min $(N_{xy}{}^*, N_{xy}')$.

For plates with aspect ratios $a/b < 2$, and for plates with clamped edges there are no simple formulae for the buckling load. In these cases equation [6.2] gives a conservative value for design, and can be used to estimate whether shear buckling is likely to be a possible failure mode.

6.2.1.3 Further references
For further general reading on panel buckling, see refs [6.1], [6.2], [6.4] and [6.6]. Specific information on rectangular orthotropic panels is given in refs [6.7]–[6.9]. PC programs for calculating panel buckling loads for different edge and load conditions are available with refs [6.7] and [6.8].

6.2.2 Transverse loads

See Fig. 6.5.

Geometry
Flat rectangular panel, sides a, b, thickness h. Edges, simply supported, clamped or free.

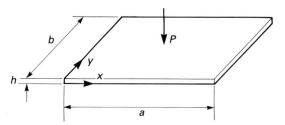

6.5 Panel under transverse load.

Applied loads
Transverse point load P, or uniform pressure p equivalent to a transverse load $P = abp$.

Design formulae
The maximum transverse deflection w in the panel is:

$$w = \alpha P a^2 / D_2 \qquad [6.13]$$

The maximum panel bending moments M_x, M_y are defined as:

$$M_x = \beta_1 P, \ M_y = \beta_2 P \qquad [6.14]$$

6.2.2.1 Design parameters, α, β_1, β_2
Panel behaviour is governed by the dimensionless stiffness parameter α and strength parameters β_1, β_2, which are functions of:

1 The plate aspect ratio a/b.
2 The material parameters D_1/D_2, D_3/D_2, v_{xy}.
3 The panel edge conditions.
4 The load type and load position.

Values of these parameters for different geometry, materials and load conditions are given from approximate design formulae, from graphs and tables computed from the orthotropic plate equations, or directly from PC programs. Some typical data sets for α, β_1, β_2 are given below.

6.2.2.2 Design procedure

Stiffness limited plates
Calculate the maximum plate deflection w from equation [6.13] using the appropriate value for α. Check that $w <$ allowed displacement.

Strength limited plates
Calculate the maximum plate bending moments M_x, M_y from equation [6.14] using the appropriate values of β_1, β_2. Check that $M_x < M_x^*$, $M_y < M_y^*$, where M_x^*, M_y^* are the plate material bending strengths (bending moments at failure). Alternatively the M_x, M_y values can be used in

CLT to check whether an appropriate biaxial failure condition has been reached.

In general, thin composite plates are *stiffness limited*, often showing very large deflections before material failure. For plates deflected so that $w/h>1$, transverse loads are carried by membrane stresses, in addition to the bending stresses discussed here. The plates are then stiffer than predicted in equation [6.13], with bending stresses lower than those given by equation [6.14]. Thus the design procedure here is conservative for plates loaded to large deflection.

6.2.2.3 Pressure loaded plates – design parameters

For plates reinforced with predominantly unidirectional (UD) fibres ($D_3/D_2 \approx 1$), the main material parameter is the orthotropy ratio D_1/D_2. Table 6.1 shows the influence of D_1/D_2 and plate aspect ratio a/b on α, β_1, β_2 for *simply supported* rectangular plates.

Figure 6.6 shows the data from Table 6.1 plotted as a design chart for α, β_1, β_2. The design charts enable a rapid assessment to be made of anisotropic effects on plate behaviour, which gives insight to the designer, e.g.:

1   High aspect ratio plates $a/b>2$ are not stiffened further by additional longitudinal fibre reinforcement.

Table 6.1 Stiffness and stress parameters for rectangular orthotropic plates under uniform pressure loading, with simply supported edges ($D_3/D_2 = 1$, $v_{\chi v} = 0.3$)

D_1/D_2		a/b				
		0.25	0.5	1	2	4
1	α	0.00321	0.00504	0.00406	0.00127	0.0002
	β_1	0.031	0.051	0.048	0.023	0.011
	β_2	0.011	0.023	0.048	0.051	0.031
2	α	0.00163	0.00303	0.0032	0.00122	0.0002
	β_1	0.031	0.059	0.069	0.033	0.014
	β_2	0.0073	0.015	0.038	0.05	0.031
4	α	0.00083	0.00163	0.00232	0.00117	0.0002
	β_1	0.032	0.063	0.094	0.052	0.02
	β_2	0.005	0.0098	0.026	0.047	0.032
7	α	0.00046	0.00093	0.00158	0.00103	0.0002
	β_1	0.032	0.064	0.112	0.078	0.026
	β_2	0.0035	0.0073	0.017	0.042	0.032
10	α	0.00033	0.00067	0.00123	0.00097	0.0002
	β_1	0.032	0.064	0.012	0.098	0.032
	β_2	0.0032	0.0062	0.014	0.039	0.032
15	α	0.00022	0.00045	0.00088	0.00083	0.0002
	β_1	0.032	0.064	0.128	0.125	0.039
	β_2	0.0023	0.0048	0.011	0.034	0.032

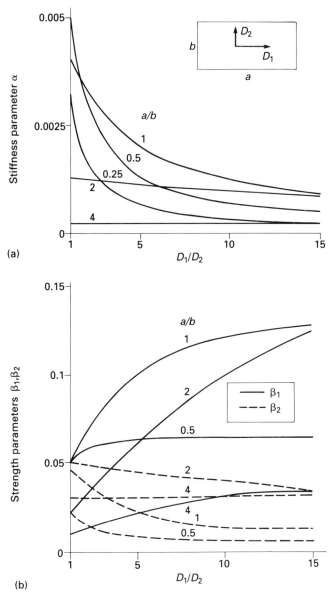

6.6 (a) Stiffness parameter α and (b) strength parameters β_1, β_2 for simply supported rectangular plates under uniform pressure loads $(D_3/D_2 = 1, v_{xy} = 0.3)$.

2   Transverse fibre reinforcement is most effective for such plates (see curves $a/b = 0.5, 0.25$).

3   The use of high performance fibres ($D_1/D_2 > 10$) does not lead to a corresponding increase in plate stiffness and strength.

4   The use of lower performance fibres ($1 < D_1 D_2 < 5$) gives a proportionately higher improvement in plate properties (i.e. design curves are steepest in this range).

Approximate design formulae
The design charts and tables were computed from numerical solutions to the orthotropic plate equations.[6.10] For simply supported rectangular plates with moderate aspect ratios $0.5 < a/b < 2$ the design parameters can be estimated by the approximate design formulae obtained from variational methods:[6.11]

$$\alpha = \quad n_1 (a/b) [D_1/D_2 + 2 (a/b)^2 (D_3/D_2) + (a/b)^4]^{-1}$$

[6.15]

$$\beta_1 = \quad n_2 \alpha [D_1/D_2 + v_{xy} (a/b)^2]$$
$$\beta_2 = \quad n_2 \alpha [v_{xy} + (a/b)^2]$$

where for *pressure*-loaded plates $n_1 = 0.0165$ and $n_2 = 9.6$.

These formulae are useful for assessing the influence of the second material parameter D_3/D_2, which can be significant in fabric reinforced and cross-ply laminates where $D_1/D_2 \simeq 1$.

6.2.2.4 Point loaded plates – design parameters

Table 6.2 shows computed values of the design parameters α, β_1, β_2 for centre point loaded rectangular plates with predominantly UD fibre reinforcement ($D_3/D_2 \simeq 1$), as functions of aspect ratio a/b and orthotropy ratio D_1/D_2.

For point loads the stress concentrations β_1, β_2 depend on the loading diameter d, which is assumed to be small. The table is based on computed solutions [6.10] in which $d = b/32$. For other loading diameters, values of β_1, β_2 can be estimated by multiplying the value in the table by the factor:

$$0.19 (0.435 + 1.3 \log 4b/\pi d)$$

[6.16]

The approximate design formula [6.15] is also valid for simply supported centre point loaded plates, with the parameter $n_1 = 0.0403$.

6.2.2.5 Further references

General design information on orthotropic plates under transverse load is given in refs [6.1], [6.2] and [6.7]. An extensive set of design tables and charts showing the influence of different edge and loading conditions, together with procedures for estimating large deflection effects is given in ref. [6.12]. A PC program PANDA, based on com-

Table 6.2 Stiffness and stress parameters for rectangular orthotropic plates under centre point loading, with simply supported edges ($D_3/D_2 = 1$, $v_{xy} = 0.3$)

D_1/D_2		0.25	0.5	1	2	4
				a/b		
1	α	0.0172	0.0166	0.0116	0.00417	0.00108
	β_1	0.397	0.445	0.435	0.365	0.297
	β_2	0.295	0.365	0.435	0.445	0.397
2	α	0.0111	0.011	0.00933	0.00378	0.00098
	β_1	0.467	0.535	0.553	0.458	0.368
	β_2	0.23	0.29	0.363	0.395	0.357
4	α	0.00707	0.00701	0.00674	0.00337	0.00088
	β_1	0.548	0.635	0.692	0.587	0.463
	β_2	0.178	0.232	0.293	0.345	0.318
7	α	0.00486	0.0048	0.00485	0.00298	0.0008
	β_1	0.62	0.725	0.813	0.725	0.558
	β_2	0.147	0.192	0.243	0.307	0.288
10	α	0.00383	0.00377	0.00384	0.0027	0.00075
	β_1	0.67	0.787	0.892	0.892	0.63
	β_2	0.128	0.17	0.217	0.282	0.272
15	α	0.0029	0.00285	0.00292	0.00234	0.0007
	β_1	0.73	0.862	0.983	0.817	0.575
	β_2	0.11	0.148	0.188	0.213	0.213

puted solutions to the orthotropic plate equations, enabling a wide range of plate loading conditions and edge conditions to be analysed, is available with ref. [6.12]. The PC program LAMPCAL, available with ref. [6.7], can be used for the analysis of simply supported rectangular composite plates under pressure or point loads. It also includes an integrated laminate analysis for the determination of the plate flexural rigidity parameters D_1, D_2 and D_3. Additional design sheets and PC programs for orthotropic plates under more complex loads are described in refs [6.13] and [6.14]. These include the calculation of plate natural frequencies[6.13] and stress concentrations in plates with circular holes.[6.14]

6.2.3 Panel design tips

1 Composite panel design is usually stiffness-limited, with panels having excellent strength properties:
   (a) in contrast metal panels are usually stiff with design controlled by the yield stress;
   (b) thin low stiffness panels may buckle under in-plane compression and shear loads;
   (c) under transverse loads very large deflections may be achieved before material fracture;

(d) in applications where failure criteria are based on the lower material cracking stresses (rather than ultimate strength), design could become stress-limited, e.g. in a tank panel where weepage must be avoided.

2 Panel geometry and edge conditions often have more influence on panel behaviour than the material properties:
   (a) clamped panels are about twice as stiff in flexure as simply supported panels, and have much higher buckling loads;
   (b) panel flexural stiffness increases as the cube of the panel thickness.

3 The correct choice of fibre orientation in composite panels gives improved panel properties:
   (a) $\pm 45°$ fibres improve the in-plane shear stiffness and strength behaviour;
   (b) in high aspect ratio panels, transverse fibres (parallel to the short side) inhibit compression buckling, and improve panel flexural stiffness and strength.

4 Thin composite panels are not very efficient for transverse load applications. Flexural stiffness is increased considerably in:
   (a) composite panels with moulded-in ribs and stringers;
   (b) sandwich panels with thin load-bearing composite skins bonded to a thick lightweight foam or honeycomb core (see Section 6.4).

6.3 Design of thin-walled beam structures

The aim of the design procedure is to calculate deflections and failure loads in a composite beam structure under defined load conditions. The design procedure, which is detailed in this section, can be summarised in the following steps:

1 Calculate beam section properties, based on material lay-up and section geometry, using the Design sheets for section properties given in Section 6.3.4.

2 Calculate beam behaviour under load with appropriate end boundary conditions. Beam design formulae for some simple load cases are given at the end of this section.

3 Calculate beam reserve factors, as discussed below, and use them to analyse beam failure behaviour and to determine maximum design load.

6.3.1 Section properties

Figure 6.7 shows a topical thin-walled composite beam structure, assumed here to be composed of thin orthotropic plate elements. Material axes (x,y) are chosen in the plane of the element, along the

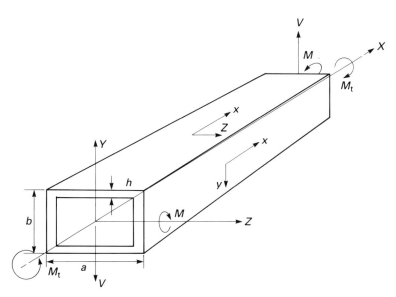

6.7 Beam co-ordinate systems and load conventions.

orthotropic symmetry axes, and the plate element is characterised by elastic stiffnesses (A), (D), moduli $E_{xx},....$, and averaged failure stresses $\sigma_{xx}^*,....$, as described in Section 6.1.2.

Axes are chosen in the beam so that the X-axis is along the beam axis, with the Y–Z axes in the plane of the cross-section, with the Y-axis vertical as shown schematically in Fig. 6.7. The X–Z plane is assumed to lie through the section's neutral axis. The x- and y-axis of material's symmetry in the composite material are defined as follows. The x-axis is always along the beam length, i.e. parallel to the X-axis, whilst the y-axis is transverse to the beam, either in the Y- or Z-direction, depending on the wall direction in the beam (see Fig. 6.7). Note that the design procedure is only valid for composite beams in which one material symmetry direction is along the beam length.

The beam is subjected to bending moments M about the Z-axis or transverse loads V in the Y-direction, both of which lead to bending about the Z-axis. Twisting moments M_t are applied about the X-axis. The convention that anti-clockwise moments are positive is adopted. With this sign convention (see Fig. 6.7) a positive bending moment, i.e. $M>0$, causes a beam to deflect so that it is concave upwards. This will often correspond to an applied load P applied downward ($-Y$ direction) and deflections w in the $-Y$ direction. As a result the longitudinal stresses in the beam are compressive on the upper face ($Y>0$) and tensile on the downward face ($Y<0$) at positions in the beam where $M>0$.

Under application of these loads, bending, transverse shear and torsion of the beam may occur. The beam section properties required for the design analysis are:

1 Stiffness properties:
   (a) flexural rigidity D;
   (b) transverse shear stiffness Q;
   (c) torsional rigidity T.
2 Strength properties:
   (a) bending moment at failure M^*;
   (b) shear force at failure V^*;
   (c) twisting moment at failure M_t^*.

The beam section properties are functions of the composite material moduli and strengths and of section geometry. Data sheets for calculating them for beams of different cross-section shapes are given in Sections 6.3.2 – 6.3.4. A thin-walled composite section may fail in a number of modes; for example, under flexural loads failure may occur by tensile or compression face material fracture, or by local buckling of a compression element. These possibilities are included in the design procedure by incorporating buckling factors ϕ into the definition of the section strength properties, which reduce the beam strength when buckling takes place.

6.3.2 Design procedure

The beam deformation when subjected to transverse loads, bending or twisting moments consists of a bending deflection w of the beam neutral axis in the vertical Y-direction, and an angle of twist θ about the beam axis. In addition there will be a distribution of bending moments, shear force and twisting moment along the beam length. The beam equations which govern the deformations, moment and force distributions in the beam will generally be the same as those for corresponding isotropic beams. However, for many composite beams a shear correction will be required in the beam analysis which takes account of the additional transverse deformations due to the low transverse shear stiffness Q of composite sections. Some basic design formulae for beams under a range of load conditions, three-point bending, end loaded cantilever, etc., are given in section 6.3.3.

From design formulae or suitable PC programs the maximum values for each load case may be calculated along the beam for:
   deflection w
   bending moment M
   shear force V
   angle of twist θ
   twisting moment M_t

By comparing these maximum values respectively with allowable deflections and twist, and the strength values for the section M^*, V^* and M_t^* it is possible for the designer to examine the failure behaviour of the beam. The beam failure mode, as well as the failure load, may be determined in the following way.

Suppose that w^* is a maximum allowable deflection in the beam and θ^* is a maximum allowable twist. Define the reserve factors:

$$R_1 = w/w^*, \ R_2 = \theta/\theta^*, \ R_3 = V/V^*, \ R_4 = M/M^*, \ R_5 = M_t/M_t^* \qquad [6.17]$$

where w, M, V, θ, M_t are the maximum values calculated from the beam analysis for a given set of applied loads. The factors R_3–R_5 are reserve strength factors, with R_1, R_2 being corresponding factors for failure by excessive deflections.

If $R_1,..., R_5 < 1$, the beam does not fail under the applied loads. On increasing the applied loads, $w, M, ...,$ will increase so that a load will be reached when one of the R-factors, say $R_i \geqslant 1$. Then failure occurs and the mode of failure corresponds to the value of i. For example, $i = 2$ implies failure by an excessive twist angle θ under torsion loads. Thus whatever the applied load, the maximum value R_i, as defined in equation [6.17], determines the first failure mode. When for a given applied load set:

$$\max \{R_i\} \begin{cases} <1 & \text{beam does not fail} \\ =1 & \text{beam just fails} \\ >1 & \text{beam has failed} \end{cases}$$

It follows that the $\max\{R_i\}$ value may be used to scale the applied loads to determine the value when failure just occurs, i.e. when $\max\{R_i\} = 1$.

By studying the calculated section failure properties, it is possible to obtain further information about the failure mode. For example, when R_4 is the controlling reserve factor, failure is by bending. However, the section bending strength M^* may refer to a material strength failure, or when the relevant buckling factor $\phi < 1$, to failure by local buckling in the section wall. Thus the design procedure allows a systematic analysis of composite beam failure properties.

Note that for this simplified failure analysis it is assumed that any interaction between bending, transverse shear and torsion loads does not modify the failure behaviour. This will not always be the case for composite beams, particularly for the onset of buckling. Similarly the principal stress material's failure criterion is assumed for the composite material, which also neglects interactions due to multiaxial loads.

6.3.3 Beam design formulae

In many loading situations standard design formulae for maximum deflections, bending moments and shear forces in isotropic beam structures may be used for composite beams; see, for example, the extensive lists of design formulae in ref. [6.4]. However, because of the low material shear properties, a shear correction to the maximum deflection may be needed for composite beams, particularly for short beams or deep sections. In these cases modified beam deflection formulae, such as those given here, may be required.

6.3.3.1 Shear correction in beams under transverse load

For a beam of length l subjected to a transverse point load P or total distributed load P, the maximum transverse deflection w is given by:

$$w = (\alpha \, Pl^3/D) \, (1 + \gamma D/l^2 Q) \qquad [6.18]$$

where the parameters α, γ depend on beam load and boundary conditions. The second bracketed quantity in equation [6.18] is a shear correction factor, which tends to unity for sections with high shear stiffness (Q large) or in long beams (l large).

Table 6.3 lists values for α, γ for several beam load conditions, with corresponding values of the maximum bending moment M and transverse shear V in the beam section.

6.3.3.2 Beams under torsion loads

See Fig. 6.8. A beam of length l is fixed at one end, with a twisting moment M_t applied to the free end. The maximum angle of twist θ at the twisted end is given by:

$$\theta = M_t l/T \qquad [6.19]$$

This equation is valid for composite beams, assuming that both ends are free to warp. By warping it is meant that the beam cross-section does not remain plane, but may deform axially. These effects are not important in compact sections, but can be significant in thin walled open sections. Young[6.4] discusses a correction method developed for isotropic beams which, in the absence of an exact analysis, may be applied also to composite beams. When warping at the ends is prevented, for example by clamping, the beam is stiffer in torsion so that the angle of twist θ is lower than predicted by equation [6.19]. Thus equation [6.19] is conservative for design purposes, and the warping correction for thin-walled open channels or an I-beam will only be required in critical situations.

If the maximum allowable angle of twist in the beam is θ^*, it follows from equation [6.19] that the maximum allowable twisting moment in the beam M_T is:

$$M_T = \min \, (M_t^*, \, T\theta^*/l) \qquad [6.20]$$

6.3.4 Design sheets for section properties

6.3.4.1 Circular tubes
See Fig. 6.9.

Table 6.3 Beam design parameters

Load conditions	α	γ	Max. bending moment M	Max. shear V
P (simply supported, central load)	1/48	12	$Pl/4$	$P/2$
P (simply supported, distributed load)	5/384	48/5	$Pl/8$	$P/2$
P (clamped, central load)	1/192	48	$Pl/8$	$P/2$
P (clamped, distributed load)	1/384	48	$Pl/12$	$P/2$
P (clamped cantilever, end load)	1/3	3	Pl	P
P (clamped cantilever, distributed load)	1/8	4	$Pl/2$	P

$\wedge$ simply supported, ■ clamped.

Geometry

Hollow tube with outer radius r, wall thickness h.
Material symmetry axes, x-axial, y-circumferential.
Neutral axis AB through centre.
Section moment of inertia about neutral axis:

$$I = \frac{\pi}{4} [r^4 - (r-h)^4] \tag{6.21}$$

for thin-walled tubes:

$$I = \pi r^3 h (1 - 3h/2r) \tag{6.22}$$

6.8 Beam under torsion load.

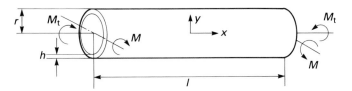

6.9 Circular tube in bending and torsion.

Stiffness properties
  Flexure: flexural rigidity for bending about *AB*:

$$D = E_{xx}I \qquad [6.23]$$

  Shear: transverse shear stiffness for thin-walled tubes:

$$Q = \frac{2}{3}\,\pi\,(2rh - h^2)G_{xy} \qquad [6.24]$$

  Torsion: torsional rigidity:

$$T = G_{xy}\,\frac{\pi}{2}\,[r^4 - (r-h)^4] \qquad [6.25]$$

Strength properties
  Flexure: The beam bending moment M^* at failure is:

$$M^* = I\phi_1\sigma_{xx}^*/r \qquad [6.26]$$

  where ϕ_1 is the buckling coefficient defined as:

$$\phi_1 = 0.65K_3\,(h/r)\,(E_{xx}/\sigma_{xx}^*)\,/(3\mu)^{1/2} \qquad [6.27]$$

  if this is <1 otherwise $\phi_1 = 1$, when no buckling occurs.
  Here:

$$K_3 = \{s\,(1 + v_{xy}s)\,[(D_3/D_1) - v_{xy}\,s^2]/\mu\}^{1/2} \qquad [6.28]$$

  where $s = (D_2/D_1)^{1/2}$.
  Shear: the transverse shear force at failure is:

$$V^* = 3I\,\sigma_{xy}^*\,[r^2 + (r-h)\,(2r-h)]^{-1} \qquad [6.29]$$

  which for thin-walled tubes reduces to:

$$V^* = \pi rh\,\sigma_{xy}^* \qquad [6.30]$$

  Torsion: the twisting moment at failure is:

$$M_t^* = 2I\,\sigma_{xy}^*/r \qquad [6.31]$$

6.3.4.2 Box beams

Geometry
    Figure 6.7 shows the box beam geometry and applied loads.
    Beam width a, depth b, wall thickness h.
    Material symmetry axes, x-axial, y-circumferential.
    Neutral axis AB through centroid along Z-axis.
    Section moment of inertia about AB:

$$I = \frac{1}{12} [ab^3 - (a-2h)(b-2h)^3] \qquad [6.32]$$

    for thin-walled sections:

$$I = \frac{1}{6} b^3 h (1 + 3a/b) \qquad [6.33]$$

Stiffness properties
    Flexure: flexural rigidity for bending about AB:

$$D = E_{xx}I \qquad [6.34]$$

    Shear: transverse shear stiffness. For thin-walled sections this
    approximates to:

$$Q = 2h (b-2h) G_{xy} \qquad [6.35]$$

    Torsion: torsional rigidity:

$$T = 2h (a-h)^2 (b-h)^2 G_{xy}/(a + b-2h) \qquad [6.36]$$

Strength properties
    Flexure: the beam-bending moment M^* at failure of the tensile/compressive faces is:

$$M^* = 2I \, \phi_2 \, \sigma_{xx}^* /b \qquad [6.37]$$

    where ϕ_2 is the compressive face buckling coefficient defined as:

$$\phi_2 = \frac{1}{3} \pi^2 K_1 (h/a)^2 (E_{xx}/\mu\sigma_{xx}^*) \qquad [6.38]$$

    if this is <1, otherwise $\phi_2 = 1$, in which case no buckling occurs.
    In the above relation K_1 takes the value defined in [6.9], i.e.:

$$K_1 = \tfrac{1}{2} [(D_2/D_1)^{1/2} + D_3/D_1)]$$

    Shear: the transverse force V^* at shear failure of the side walls is:

$$V^* = \frac{4}{3} bh\psi \, \sigma_{xy}^* \qquad [6.39]$$

where ψ is the side wall shear buckling coefficient:

$$\psi = 6.6K_2 \, (h/b)^2 \, (E_{xx}/\mu\sigma_{xy}{}^*) \qquad [6.40]$$

if this is <1, otherwise $\psi = 1$, when no buckling occurs. Here K_2 is as defined in equation [6.12].

Torsion: the twisting moment at failure:

$$M_t{}^* = 2hab \, \sigma_{xy}{}^* \qquad [6.41]$$

6.3.4.3 U-sections
A U-section is shown in Fig. 6.10.

Geometry
Base width a, thickness h.
Rib side wall height e, thickness $\frac{1}{2}d$.

Rib height ratio, $H = e/h$.
Rib thickness/base width ratio, $W = d/a$.
Neutral axis AB at distance y_1 from base:

$$y_1 = \frac{ah^2 + de \, (2h + e)}{2 \, (ha + de)} \qquad [6.42]$$

Section moment of inertia about AB:

$$I = \frac{1}{12} \, a \, h^3 S_1 \qquad [6.43]$$

where the rib stiffness factor S_1, which defines the increased rigidity of the section relative to the base plate, is given by:

$$S_1 = (1 + 4WH + 6WH^2 + 4WH^3 + W^2H^4) \, / \, (1 + WH) \quad [6.44]$$

Stiffness properties
Flexure: flexural rigidity for bending about AB:

$$D = E_{xx}I \qquad [6.45]$$

Shear: transverse shear stiffness. For deep sections, i.e. large e/h values, Q is given approximately by:

$$Q = \frac{5}{6} \, de \, G_{xy} \qquad [6.46]$$

Torsion: torsional rigidity:

$$T = \frac{1}{3} \, (ah^3 + \frac{1}{4} \, ed^3) \, G_{xy} \qquad [6.47]$$

For thin-walled U-sections warping restraint can be very significant, and should be checked for clamped end conditions, see [6.4, section 9.3].

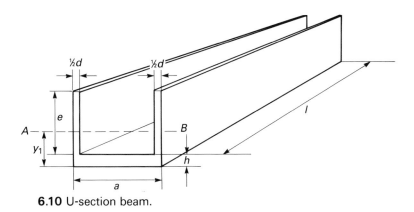

6.10 U-section beam.

Strength properties

  Flexure: for this non-symmetrical section several failure modes are possible and it is necessary to distinguish between the cases when the ribs are loaded in tension and those loaded in compression. For ribs loaded in *compression* where failure may occur of the ribs in compression or of the face in tension, the bending strength M^* is defined as:

$$M^* = \min \{I\sigma^*_{xx}/y_1, I\varphi_3\sigma^*_{xx}/(b-y_1)\} \qquad [6.48]$$

  where the side wall buckling factor φ_3 is:

$$\varphi_3 = 0.35K_1(d/2e)^2(E_{xx}/\mu\sigma^*_{xx}) \qquad [6.49]$$

  if this is <1, otherwise $\varphi_3 = 1$ when no buckling occurs. The parameter K_1 is defined in equation [6.9].

  For ribs loaded in *tension* where failure may occur in the ribs in tension or as a compression face failure, the bending strength M^* is given by:

$$M^* = \min\{I\sigma^*_{xx}/(b-y_1), I\phi_2\sigma^*_{xx}/y_1\} \qquad [6.50]$$

  where ϕ_2 is defined in equation [6.38].

  Note that for composite materials the material tensile and compressive strengths are often different. It may be necessary to use the appropriate value for the failure mode when using equations [6.48]–[6.50].

Shear: the transverse shear force V^* at failure due to material shear failure in the section is:

$$V^* = 0.67ahS_1 \, \sigma^*_{xy}(1+WH^2)/(1+2H+2WH+3WH^2)^2 \qquad [6.51]$$

Torsion: there are no simple formulae for the torsion strength of a U-section. Table 20 in Young[6.4] discusses such cases with re-entrant corners and gives an approximate formula. Since composite

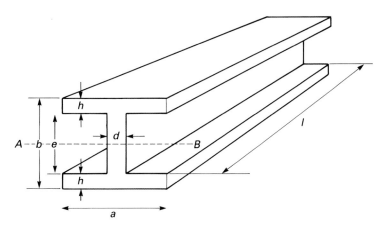

6.11 I-section beam.

beams usually have relatively low torsional stiffness, failure occurs usually by excessive torsional deflection rather than through a material shear stress failure. Thus formulae for the maximum twisting moment M_t^* are not given here.

6.3.4.4 I-sections

An I-section is shown in Fig. 6.11.

Geometry
Flange width a, thickness h.
Beam depth b, web depth e, web thickness d.
Width ratio, $W = d/a$.
Height ratio, $H = e/2h$.
Neutral axis AB through centroid.
Section moment of inertia about AB:

$$I = \frac{1}{12}\left[ab^3 - (a-d)(b-2h)^3\right] \tag{6.52}$$

for thin-walled sections:

$$I = \frac{1}{12} b^3 (d + 6ha/b) \tag{6.53}$$

Stiffness properties
Flexure: flexural rigidity for bending about AB:

$$D = E_{xx}I \tag{6.54}$$

Shear: transverse shear stiffness. For thin-walled sections this approximates to:

$$Q = deG_{xy} \tag{6.55}$$

Torsion: torsional rigidity:

$$T = \frac{1}{3}(2ah^3 + ed^3)G_{xy}$$

[6.56]

for thin-walled sections. Warping restraint can be important in I-beams with clamped ends, see ref. [6.4, Section 9.3].

Strength properties
Flexure: the bending moment at failure in the flanges M^* is:

$$M^* = 2I\varphi_4\sigma^*_{xx}/b$$

[6.57]

where φ_4 is given by:

$$\varphi_4 = 1.68K_1(h/a)^2(E_{xx}/\mu\sigma^*_{xx})$$

[6.58]

if this is < 1, otherwise $\varphi_4 = 1$.
Shear: the transverse force V^* at shear failure in the web is:

$$V^* = \frac{2}{3}ed\,\psi\sigma^*_{xy}$$

[6.59]

where ψ is the web shear buckling coefficient given by:

$$\psi = 6.6K_2\,(d/e)^2\,(E_{xx}/\mu\sigma^*_{xx})$$

[6.60]

if this is < 1, otherwise $\psi = 1$, when no buckling occurs. Here K_2 is defined by equation [6.12].
Torsion: there is no simple formula for the torsion strength of an I-section, see Table 20.1 in Young.[6.4]

6.3.5 Beam design tips

The design procedures given above may be used for calculating composite beam behaviour under load. However, in many situations the designer must first select the material lay-up and exact geometry (see Fig. 6.1), so that the design formulae cannot be used directly. In such a case it is advantageous to use the design formulae to produce design charts for understanding structural behaviour and to aid in materials' selection. The required design charts must be tailored to a particular application. Two examples of this approach to design are given here.

6.3.5.1 Bending and torsion of a composite tube

Figure 6.12 shows an FRP tube fabricated by filament winding in which fibres are wound at angles $\pm\,\theta°$ to the tube axis. Such a tube could be used for a vehicle drive shaft or as an aircraft control lever, where the flexural and torsional stiffnesses of the tube are required. For the composite tube these properties can be tailored to the application by choice of angle θ.

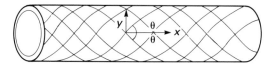

6.12 Filament wound cylinder.

From equations [6.21], [6.23] and [6.24] the ratio of the flexural and torsional rigidities is:

$$D/T = E_{xx}/2G_{xy}$$

which is independent of tube diameter and thickness and depends only on the material properties. For isotropic materials, such as metals:

$$D/T = E/2G = 1 + v$$

which is approximately constant, showing that it is not possible for both the bending and torsion stiffnesses to be varied independently.

For the composite tube both E_{xx} and G_{xy} are functions of the winding angle θ and of the elastic constants of a UD ply E_{11}, E_{22}, G_{12}, v_{12} with respect to the (1,2) materials axes in the ply. Here the 1-axis is in the fibre direction and the 2-axis transverse to the fibres. Using design formulae for an angle-ply laminate (see for example equation [5.19] and [5.21]) or by using a PC laminate analysis program the dependence of E_{xx} and G_{xy} on θ can be easily calculated. The ratio D/T is then a function of winding angle θ, and through the correct choice of θ a tube with required properties can be designed.

The determination of θ is most easily carried out by means of the design chart (Fig. 6.13). This shows the flexural and torsional rigidities normalised by $E_{11}I$, the flexural rigidity of the 0° tube, as a function of winding angle θ for a glass reinforced plastic (GRP) tube with the following typical material property ratios:

$$E_{22}/E_{11} = 0.22, \ G_{12}/E_{11} = 0.11, \ v_{12} = 0.26$$

The design chart shows that the ± 45° GRP tube has the maximum torsional rigidity, the 0° tube the maximum flexural rigidity and that a tube with $\theta = \pm 34°$ would have equal torsional and flexural rigidities.

6.3.5.2 Buckling failures in box-beams
The onset of failure in thin-walled composite box beams is often initiated by compression face buckling and, as described in section 6.3.3, this is controlled by the buckling parameter ϕ_2. The design chart (Fig. 6.14) enables a rapid assessment to be made of the effect of different types of reinforcement on the stability of a box-beam. Figure 6.14 shows the buckling factor ϕ_2 as a function of the parameter $(a/h)(\sigma^*_{xx}\mu/E_{xx})^{1/2}$, for different values of the material parameter K_1.

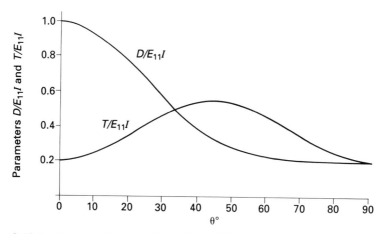

6.13 Design chart for selection of θ for GRP tube under bending and torsion ($E_{22}/E_{11} = 0.22$, $G_{12}/E_{11} = 0.11$, $v_{12} = 0.26$).

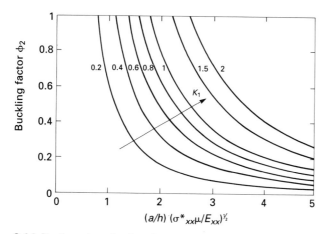

6.14 Design chart for box-beam buckling factor ϕ_2.

The function σ^*_{xx}/E_{xx} is a failure strain and is approximately constant for many composite materials and lay-ups. Thus the design chart shows ϕ_2 as a function mainly of a/h, the width/thickness of the compression face.

The parameter K_1 defined in equation [6.9] depends on the beam wall flexural rigidities. $K_1 = 1$ corresponds to an isotropic material. Figure 6.14 shows that the buckling coefficient ϕ_2 is reduced when $K_1 < 1$ and increased when $K_1 > 1$, in comparison with an isotropic material. Some typical values for a GRP box-beam with different materials' lay-ups are:

UD longitudinal	$K_1 =$	0.37
UD transverse	$K_1 =$	1.7
$\pm 45°$	$K_1 =$	1.4
$0°/90°$	$K_1 =$	0.72

We see that transverse fibres in the beam tend to stabilise against side-wall buckling since $K_1 = 1.7$, whilst longitudinal fibres where $K_1 = 0.37$ lead to an increased probability of a buckling failure. Thus the design chart may be used as an aid in materials' selection and as a guide in controlling the failure mode. Since $\phi_2 \geqslant 1$ corresponds to a material fracture rather than a buckling failure in the beam, for a given K_1 value the chart can be used to define the critical ratio a/h at which $\phi_2 = 1$, which is the boundary point between a local buckling failure and material fracture.

6.3.5.3 General beam design tips

1 Beam structures with longitudinal fibres have the highest bending stiffness and strength.
2 Beams with $\pm 45°$ fibres usually have the highest torsional stiffness and strength.
3 In thin-walled beams local wall buckling may occur which leads to premature collapse prior to material fracture.
4 In general, longitudinal fibre reinforcement increases the likelihood of local buckling, whereas transverse fibre reinforcement stabilises the beam structure.
5 Buckling failures can be effectively reduced by corner reinforcements, e.g. by triangular fillets or by internally radiused corners.
6 It follows from (1)–(4) that for many practical applications composite beam structures will require both longitudinal and transverse or $\pm$ 45° fibres. Thus final beam design is strongly dependent on the actual load conditions and permitted failure modes.
7 For shorter composite beams with length/thickness less than about 10, transverse shear effects can no longer be neglected and shear corrections may be required in the design analysis (see Section 6.4.2).

6.3.5.4 Further references

Extensive information on beam design formulae for isotropic materials is given in ref. [6.4], and much of this information can be adapted for composite beams provided that appropriate materials' moduli and strength values are used. Design procedures for specific beam sections are discussed in refs [6.5], [6.15]–[6.17]. The PC program MIC-MAC[6.5] has modules for the analysis of composite beams with circular, elliptic and box sections under bending and torsion load. However, buckling failures are excluded. Buckling failures of a circular composite cylinder under axial compression load and under torsion load can be analysed with the PC program CYLAN.[6.18] Tests on GRP box and open sections which validate some of the design procedures described here can be found in refs [6.19] and [6.20].

6.4 Design of sandwich structures

Sandwich materials consisting of thin FRP skins bonded onto a thick low density core are widely used for lightweight constructions where the main loads are flexural. Typical high performance sandwich materials for aerospace applications or for racing car chassis consist of carbon fibre–epoxy laminated skins, with aluminium honeycomb or resin impregnated aramid fibre paper honeycomb; see Fig. 6.15. For application in vehicles, boats, building panels, glass fibre fabric or mat reinforced skins are used with PU and PVC foam cores, or wood cores of end-grain balsa and plywood.

Sandwich construction is particularly efficient under flexural loads since the stronger skins are at the surface where flexure stresses are high, and they take the main loads in tension or compression. The core, which is a low stress region, consists of a low weight, low modulus material which carries the transverse shear loads. It also acts as a spacer for the skins, thus increasing the plate thickness, second moment of area and hence flexural rigidity. The core may also provide thermal insulation, energy absorption properties, etc., to the panel.

This subsection gives only an introduction to sandwich construction, consisting of approximate formulae for determining sandwich materials stiffness and strength properties and failure modes, which are suitable for the preliminary design of a sandwich lay-up. This is followed by a discussion of how design procedures for plates and beams discussed in Sections 6.2 and 6.3 require modification for sandwich materials.

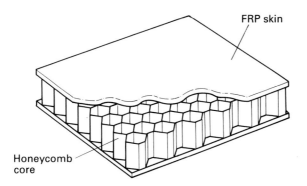

6.15 Sandwich panel with honeycomb core.

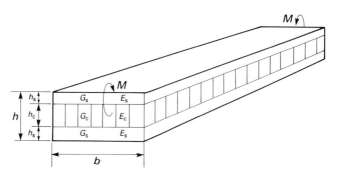

6.16 Schematic diagram of sandwich beam element.

6.4.1 Properties of sandwich materials

Figure 6.16 shows an element of a typical symmetric sandwich material, which could be a sandwich beam or an element from a sandwich panel. The sandwich geometry is defined by the skin and core thicknesses h_s, h_c. For simplicity the formulae presented here apply to isotropic materials, but they are also valid for orthotropic materials, in which case the moduli and strength values would be given appropriate values for the panel x- or y-direction.

The required beam element properties, for the beam of width b, are:

1 Stiffness properties:
    (a)   in-plane stiffness A;
    (b)   flexural rigidity D;
    (c)   transverse shear stiffness Q.
2 Strength properties:
    (a)   in-plane failure load N^*;
    (b)   bending moment at failure M^*;
    (c)   transverse shear stiffness V^*.

These properties are functions of:

1 Sandwich element geometry:
    (a) h_s, h_c – skin and core thickness;
    (b) $h = h_c + 2h_s$ – sandwich thickness;
    (c) $u = h_c/h$ – core/sandwich thickness ratio;
    (d) l, b – element length, width.
2 Skin material properties:
    (a) E_s, σ_s^* – tensile modulus and strength;
    (b) G_s, τ_s – in-plane shear modulus and strength.
3 Core material properties:
    (a) E_c, σ_c^* – tensile or compressive modulus and strength;
    (b) G_c, τ_c – through-thickness shear modulus and strength.

6.4.1.1 Sandwich stiffness properties

For the symmetric sandwich element (Fig. 6.16) under uniaxial load, we have the following approximate design formulae.[6.21]

Tensile stiffness A, effective tensile modulus E_T:

$$A = bhE_T$$
$$E_T = E_s(l-u) + E_c u \qquad [6.61]$$

where $u = h_c/h$ is the core/sandwich thickness ratio.

Flexural rigidity D, effective flexural modulus E_B:

$$D = \frac{1}{12} bh^3 E_B$$
$$E_B = E_s(1-u^3) + E_c u^3 \qquad [6.62]$$

Transverse shear stiffness:

$$Q = bh_c G_c \qquad [6.63]$$

where it is assumed that all the shear stresses are transmitted through the core.

For an idealised sandwich with thin, stiff skins and a thick, low modulus core $h_s \ll h_c, h, E_c \ll E_s$ and equations [6.61] and [6.62] may be approximated to:

$$A = 2bh_s E_s$$
$$D = \frac{1}{2} bh_s h_c^2 E_s \qquad [6.64]$$

6.4.1.2 Sandwich strength properties

Sandwich materials exhibit a number of distinctive failure modes (see Fig. 6.17), and in determining strength values it may be necessary to check more than one failure mode to determine the lowest strength value for design.

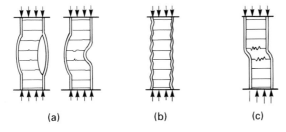

6.17 Sandwich material failure modes in compression: (a) skin wrinkling; (b) skin dimpling; (c) core crimping.

Tensile failure load N*
For *skin* failure:

$$N^* = bhE_T\sigma_s^*/E_s \qquad [6.65]$$

which for thin skin approximates to:

$$N^* = 2bh_s\sigma_s^*$$

For *core* failure:

$$N^* = bhE_T\sigma_c^*/E_s \qquad [6.66]$$

which for thin skin approximates to:

$$N^* = 2bh_sE_s\sigma_c^*/E_c$$

If the core fails first, the loads will be carried by the skins, thus equation [6.65] gives the maximum tensile load N^*.

Compressive failure load N*
Equations [6.65] and [6.66] are valid for compressive material failures in the skin and core respectively, with appropriate compression strength values. However, under compression other failure modes are possible.

Skin wrinkling (Fig. 6.17(a))
Based on buckling of a plate on an elastic foundation, the skin wrinkling stress, σ_s' is [6.22] given as:

$$\sigma_s' = \frac{1}{2}(E_sE_cG_c)^{1/3} \qquad [6.67]$$

Skin dimpling (Fig. 6.17(b))
In honeycomb or other cellular core structures the skins may wrinkle between the cell walls. This dimpling stress σ_s' depends on the cell size d.[6.23]

$$\sigma_s' = 0.75E_s(h_s/d)^{3/2} \qquad [6.68]$$

Core crimping (Fig. 6.17(c))
In this case axial compression loads in the sandwich lead to shear deformation in the core, at the approximate compressive stress.[6.23]

$$\sigma_s' = G_ch/2h_s \qquad [6.69]$$

Whenever $\sigma_s' < \sigma_s^*$, the skin material compression strength, for one of the cases [6.67]–[6.69] the lowest value of σ_s' is used in equation [6.65] to calculate the sandwich compressive failure load N^*.

Flexural strength M*
This is the sandwich bending moment at failure. Based on tensile or compressive *skin* failure:

$$M^* = 2D\sigma_s^*/hE_s \tag{6.70}$$

which for thin skins approximates to:

$$M^* = bh_sh_c\sigma_s^*$$

In flexure, skin wrinkling or dimpling are possible failure modes on the compression face. If from equations [6.67] or [6.68] $\sigma_s' < \sigma_s^*$, the lower value is used in equation [6.70] for calculating M^*.

For core failure:

$$M^* = 2D\sigma_c^*/h_cE_c \tag{6.71}$$

which for thin skins approximates to:

$$M^* = bh_sh_cE_s\sigma_c^*/E_c$$

The flexural strength is the lower value of equations [6.70] and [6.71].

Transverse shear strength V*
If it is assumed that all the shear stresses are carried in the core:

$$V^* = bh_c\tau_c \tag{6.72}$$

6.4.1.3 Design of sandwich materials

The simplified formulae listed above enable the designer to select suitable skin and core materials and thicknesses for a sandwich material with particular properties. Properties such as flexural rigidity and tensile stiffness can be calculated more precisely for a composite sandwich material using laminate analysis programs, such as ref. [6.5]. The simplified formulae, however, enable an assessment to be made of possible failure modes, which can then be prevented by the correct choice of materials and thicknesses. More sophisticated material selection procedures based on a minimum weight sandwich design are also available, see, for examples refs. [6.21], [6.22] and [6.24]. These procedures are not discussed further here, since the optimised sandwich constructions are often found to have very thin skins which are not always robust enough for practical applications.

6.4.2 Design of sandwich beams

The design analysis of sandwich beam structures follows the methods set out in Section 6.3 for composite beam structures. They are based on standard beam design formulae such as equation [6.18], together with sandwich beam properties such as flexural rigidity D and flexural strength M^* computed from the formulae of Section 6.4.1 above.

Because sandwich materials may have a low shear stiffness due to the low modulus core, particular attention has to be paid to shear effects. The assessment of shear effects is facilitated by classifying the materials according to their shear stiffness.

The maximum deflection w in a beam of length l with flexural rigidity D and shear stiffness Q under a transverse load P is given by equation [6.18]. The bracketed term $[1 + \gamma D/l^2 Q]$ is the correction factor to the deflection due to transverse shear effects. This factor tends to unity for long beams, or for beams with a high shear stiffness Q.

Sandwich materials are classified as follows:

stiff-core sandwich: $D/l^2 Q < 0.01$ shear effects small   [6.73]
soft-core sandwich: $D/l^2 Q > 0.1$ shear effects dominant [6.74]

For a beam under three-point bending, Table 6.3 gives $\gamma = 12$. Thus with the above classification, for stiff-core sandwiches the shear correction in three-point bending is <12% to the maximum deflection. For an idealised sandwich with thin skins:

$$D/l^2 Q \simeq h_c h_s E_s / 2 l^2 G_c$$

and it follows that the sandwich has a soft-core classification when the core thickness h_c is large, the core shear modulus G_c is small or when the beam length l is small.

It is thus convenient to define a critical length l^* for structural analysis, which follows from equation [6.73] as:

$$l^* = 10(D/Q)^{1/2} \tag{6.75}$$

For sandwich beams with length $l > l^*$, we have:

1 Stiff core sandwich.
2 Shear effects in bending are small.
3 Standard design based on beam bending may be used.

For sandwich beams with length $l < l^*$ we have:

1 Soft core sandwich.
2 Shear effects increase the bending deflections.
3 A shear correction based on equations such as [6.18] is needed in a bending analysis.

For all sandwich beams:

1 Shear effects are important when the characteristic length is small, for example in load introduction regions, in panel vibration or local buckling.
2 The shear strength V^* (equation [6.72]) should be compared with the maximum transverse shear load V in the structure to ensure that the core shear strength is adequate.

6.4.3 Design of sandwich panels

6.4.3.1 Properties of panel materials

The approximate materials design formulae [6.61]–[6.64] can be extended to sandwich plates with orthotropic skins and an isotropic or orthotropic core. With reference to x–y co-ordinates in the plate in the orthotropic symmetry directions, we need the skin material properties, E_{xxs}, E_{yys}, $G_s = G_{xys}$, $v_s = v_{xys}$ and the core material properties, E_c, G_c. For a symmetric sandwich plate with orthotropic skins of thickness h_s and core of thickness h_c we have the following approximate design formulae.

In-plane stiffnesses [A]

$$A_{11} = (h/\mu)[E_{xxs}(1-u)+E_cu]$$
$$A_{22} = (h/\mu)[E_{yys}(1-u)+E_cu]$$
$$A_{12} = v_sA_{22}$$
$$A_{66} = h[G_s(1-u)+G_cu] \qquad [6.76]$$

where as a simplifying assumption it has been assumed that the sandwich Poisson ratios are equal to the skin values, and that here:

$$\mu = 1-v_s^2E_{yys}/E_{xxs}$$

Flexural rigidities [D]

$$D_{11} = (h^3/12\mu)[E_{xxs}(1-u^3)+E_cu^3]$$
$$D_{22} = (h^3/12\mu)[E_{yys}(1-u^3)+E_cu^3]$$
$$D_{22} = v_sD_{22}$$
$$D_{66} = (h^3/12)[G_s(1-u^3)+G_cu^3] \qquad [6.77]$$

Transverse shear stiffness Q

$$Q = h_cG_c \qquad [6.78]$$

The plate stiffnesses [A], [D] may be calculated more precisely using CLT programs. The sandwich plate strength properties may be estimated by generalising equations [6.65]–[6.72], using appropriate moduli and strength values and invoking uniaxial failure criteria.

6.4.3.2 Sandwich plates under transverse load

In the design analysis of sandwich plates under transverse loads, attention must be paid to shear effects as discussed above for sandwich beam design. Sandwich plates may be classified similarly, thus:

1 Stiff-core sandwich:
   (a) rectangular plates with geometry, a, $b > l^*$, the critical length defined in equation [6.75];
   (b) standard plate design methods, see Section 6.2.2, are used for

determining maximum deflection w and bending moments M_x, M_y;

(c) plate strengths are established by comparing maximum bending moments with sandwich bending strengths M^*_x, M^*_y derived from equations [6.70] and [6.71];

(d) the maximum transverse shear load V in the plate is determined from the applied loads and compared with the sandwich shear strength V^* from equation [6.72] to ensure that the core does not fail in shear.

2   Soft core sandwich:

(a) rectangular plates with a, $b < l^*$;

(b) the plate design method described above is modified by a shear correction to the maximum bending deflection w.

The shear correction may be estimated from a modified form of the plate design formula [6.13]:

$$w = (\alpha Pa^2/D_2)[1+\gamma D_2/a^2Q] \qquad [6.79]$$

The parameter γ depends on load conditions, geometry a/b and material constants such as D_1/D_2, and can be determined from solutions to the orthotropic thick plate equations. An estimate of γ suitable for a first approximation to the shear correction may be obtained from the appropriate value for beams listed in Table 6.3.

Specific design charts for sandwich plates have been derived from the thick plate equations [6.23]. Figure 6.18 gives an example of such a design chart for calculating the maximum deflection w in a rectangular isotropic sandwich plate with simply supported edges under a uniform pressure load p. The non-dimensional deflection function is shown as a function of b/a for different values of the shear parameter U:

$$U = \pi^2 D/b^2Q$$

which is the plate equivalent of the beam shear parameter D/l^2Q.

The design chart enables plate deflections to be determined and an assessment of shear effects to be made. It shows that when $U < 0.1$ the shear correction is small, and it only becomes significant for $U > 0.2$.

6.4.3.3 Further reading

Detailed information on the design of sandwich materials and structures can be found in refs [6.2], [6.22] and [6.23]. In particular these publications contain design sheets for sandwich panels under transverse loads and under in-plane buckling loads, which enable shear corrections to be made. These shear corrections can be significant in plate buckling and vibration, because of the smaller characteristic length of the buckle or vibration wavelength. Relevant design charts similar to Fig. 6.18 for buckling are given in ref. [6.23]. Skin wrinkling in sandwich panels is considered in ref. [6.25] along with a PC program for

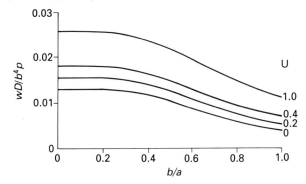

6.18 Design chart for the maximum deflection w in isotropic sandwich panels under pressure p, with simply supported edges, showing shear correction (after ref. [6.23]).

calculating wrinkling stresses. Design sheets for the natural frequencies of sandwich panels with shear corrections can be found in ref. [6.26].

6.4.4 Sandwich design tips

1 Sandwich materials are efficient structures for beams and panels under flexural loads. A sandwich panel will always have considerably higher flexural properties than an equal weight panel composed of the face plates alone.
2 In-plane tensile properties of sandwich materials are usually lower than those of the equal weight non-sandwich panel. Compression buckling loads will be higher for the sandwich because of the higher flexural rigidities.
3 An optimised sandwich construction of least weight will usually have very thin skins and a thick core. The limiting design criteria then become skin wrinkling, dimpling and local indentation effects.
4 For sandwich materials with low shear stiffness, shear corrections are required under transverse loads and for the analysis of panel buckling and vibration. This is the case for thick core materials and for short beam or panel lengths.
5 Load introduction needs careful attention in sandwich materials, since the loads have to be introduced into the thin load-bearing skins. Special fitments and edge fasteners are usually moulded or bonded into the sandwich material for this purpose.

6.5 Design of thin-walled vessels

This section gives a brief introduction to the design of cylindrical composite vessels and pipes. It concentrates on one aspect which is impor-

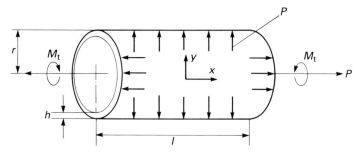

6.19 Cylindrical pressure vessel geometry and loads.

tant for preliminary design and illustrates the design freedom with composites, that is the selection of fibre orientation in filament wound cylinders. More practical aspects of the design of FRP vessels and pipes such as joints, flanges, cut-outs, etc., are not considered here.

6.5.1 Open and closed cylinders

We consider the general analysis of a thin-walled FRP cylindrical vessel (Fig. 6.19) under a combination of internal pressure, axial and torsion loads.

Geometry

> External radius r
> Length l
> Wall thickness h
> Thin-walled $h/r < 0.1$
> Open ends FRP pipe
> Closed ends FRP pressure vessel
> Orthotropic materials symmetry axes, x – axial, y – circumferential.

Load

> Internal pressure p
> Axial load P
> Torsion moment M_t

Stresses
For thin-walled vessels the stresses in the cylindrical wall are as in Ref. [6.5]:

$$\sigma_{xx} = \lambda pr/2h + P/2\pi rh$$
$$\sigma_{yy} = pr/h$$
$$\sigma_{xy} = M_t/2\pi r^2 h$$

[6.80]

where

$$\lambda = \begin{cases} 1, \text{ closed ends} \\ 0, \text{ open end} \end{cases}$$

Strains
The wall strains ε_{xx}, ε_{yy}, ε_{xy} in the material (x, y) co-ordinate system are computed direct from the stresses by substitution of equation [6.80] in [6.3].

Deformations

Axial:	$u_x = l\varepsilon_{xx}$	
Radial:	$u_r = r\varepsilon_{yy}$	[6.81]
Angle of twist:	$\alpha = l\varepsilon_{xy}/r$	

Design procedure
When the laminate construction in the vessel wall is known and hence the material properties E_{xx}, σ^*_{xx},..., the analysis procedure for a particular load case is as follows:

1 Calculate the wall stresses σ_{xx}, σ_{yy}, σ_{xy} from equation [6.80].
2 Calculate the vessel reserve strength factors R defined either by the stress ratios:

$$R = \min \left[\sigma_{xx}/\sigma^*_{xx}, \sigma_{yy}/\sigma^*_{yy}, \sigma_{xy}/\sigma^*_{xy} \right]$$

or from CLT software using the multiaxial wall stress state.
3 If $R < 1$ the structure does not fail under the load set. The actual failure load is then obtained by scaling up the applied loads by a factor $1/R$.
4 Determine the strains and hence the displacements u_x, u_r and twist α, and compare with allowable deformation criteria.

Often the laminate construction will not be known, since the designer wishes to determine this in order to optimise the vessel for a particular load set. In this case the above procedure is repeated as a design cycle with different lay-ups, or the design formulae may be used as the basis for a set of design charts, as in the design examples in the next section.

6.5.2 Filament wound vessels

A pipe under internal pressure is shown in Fig. 6.20.

Geometry

Length l, radius r, wall thickness h.
Ends open and unconstrained, $\lambda = 0$.
Filament wound with fibre directions $\pm \theta$.

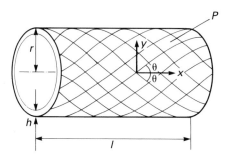

6.20 Filament wound pipe or vessel.

Loads

Internal pressure p.
$P = M_t = 0$.

The wall stresses from equation [6.80] are:

$$\sigma_{yy} = pr/h$$
$$\sigma_{xx} = \sigma_{xy} = 0 \qquad\qquad [6.82]$$

and substitution into equation [6.3] gives for the strains:

$$\varepsilon_{xx} = -v_{yx}\, pr/hE_{yy}$$
$$\varepsilon_{yy} = pr/hE_{yy}$$
$$\varepsilon_{xy} = 0. \qquad\qquad [6.83]$$

Then from equation [6.81] the displacements are:

$$u_x = -v_{yx}prl/hE_{yy}$$
$$u_r = pr^2/hE_{yy} \qquad\qquad [6.84]$$

Choice of fibre orientation θ
From equation [6.82] the pipe wall is stressed uniaxially in the circumferential or hoop direction. The burst pressure p^* is thus:

$$p^* = h\sigma^*_{yy}\,/r \qquad\qquad [6.85]$$

For the filament-wound pipe the hoop strength, σ^*_{yy} is dependent on fibre angle θ. The maximum burst pressure occurs when σ^*_{yy} takes its maximum value. For this simple uniaxial loading case this is the case when $\theta = 90°$ so that all the fibres are in the hoop direction and $\sigma^*_{yy} = \sigma^*_{11}$, where σ^*_{11} is the uniaxial strength along the fibres in a UD ply. With $\theta = 90°$ the wall moduli take the following values for a UD ply: $E_{xx} = E_{22}$, $E_{yy} = E_{11}$, $v_{yx} = v_{12}$, and hence from equations [6.84] and [6.85] the burst pressure and maximum pipe displacements are:

$$p^* = h\sigma^*_{11}/r$$
$$u_x = -v_{12}p^*rl/hE_{11}$$
$$u_r = p^*r^2/hE_{11} \qquad\qquad [6.86]$$

By referring to the earlier design chart, Fig. 6.13, it can be seen that a filament-wound cylinder with $\theta = 90°$ has the lowest bending and torsional stiffness. Thus if the pipe requires a specific bending or torsional stiffness the design procedure here will need modifying to select the fibre angle θ. For example, if an applied twisting moment M_t were known, then an analysis under combined internal pressure and axial twist could be used to determine a suitable fibre angle θ.

6.5.2.1 Vessel under pressure

Geometry

> As in Fig. 6.20, with ends closed.
> Length l, radius r, wall thickness h.
> Ends closed, $\lambda = 1$.
> Filament wound with fibre directions $\pm\,\theta$.

Loads

> Internal pressure p.
> $P = M_t = 0$.

The wall stresses from equation [6.80] are:

$$\sigma_{xx} = pr/2h$$
$$\sigma_{yy} = pr/h$$
$$\sigma_{xy} = 0 \qquad\qquad\qquad [6.87]$$

and substitution into equation [6.3] gives the strains:

$$\varepsilon_{xx} = pr\,(1/2-\nu_{xy})/hE_{xx}$$
$$\varepsilon_{yy} = pr\,(1-\nu_{yx}/2)/hE_{yy}$$
$$\varepsilon_{xy} = 0 \qquad\qquad\qquad [6.88]$$

and hence from equation [6.81] the displacements are:

$$u_x = prl\,(1/2-\nu_{xy})/hE_{xx}$$
$$u_r = pr^2\,(1-\nu_{xy}\,E_{yy}/E_{xx})/hE_{yy} \qquad\qquad [6.89]$$

Choice of fibre orientation θ
A strategy is now required for determining the fibre angle θ. In this case there are biaxial stresses in the cylindrical wall and two cases of interest are the value of θ corresponding to:

1 Maximum internal pressure at wall failure under the biaxial stresses.
2 Minimising the displacements u_x or u_r.

This is achieved by using the design formulae to produce design charts showing the influence of winding angle on vessel properties.
    Let $\sigma = pr/h$; from equation [6.87] the biaxial wall stresses are $\sigma_{xx} =$

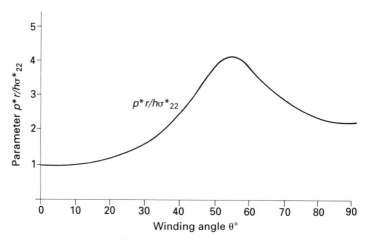

6.21 Design chart for failure pressure p^* in an a $\pm\,\theta$ wound cylinder GRP pressure vessel (E_{11} = 38.6 GPa, E_{22} = 8.3 GPa, G_{12} = 4.1 GPa, v_{12} = 0.26, σ_{11}^* = 1062 MPa, σ_{22}^* = 31 MPa, σ_{12}^* = 72 MPa).

$\sigma/2$, $\sigma_{yy} = \sigma$ and we require the strength function $\sigma^*\,(\theta)$ of a $\pm\,\theta$ angle-ply laminate under this particular biaxial stress system. This function is readily calculated using CLT software. This was carried out with data for a UD GRP material using the program LAMICALC[6.27] using a first ply failure (FPF) condition with a Tsai-Wu quadratic stress failure criterion. Since FPF corresponds here to matrix cracking transverse to the fibres in a UD ply, the calculated biaxial failure stress $\sigma^*\,(\theta)$ corresponds to weepage stress for the vessel. In terms of σ^*, the failure pressure p^* at weepage is:

$$p^* = \sigma^* h/r \qquad\qquad [6.90]$$

The computed function $\sigma^*(\theta)$ is shown in Fig. 6.21, non-dimensionalised by the UD ply transverse strength σ_{22}^*, as a design chart for p^*. Thus Fig. 6.21 may be used to determine the maximum design pressure $p^*r/h\sigma^*_{22}$ in the vessel at weepage as a function of winding angle θ. A similar design chart could be produced based on a last ply failure (LPF) stress in the angle-ply laminate to give the ultimate burst pressure p^* in the vessel. The design chart shows that the optimum winding angle for the GRP vessel is about $\theta = \pm\,55°$. This corresponds also to the optimum winding angle obtained from a simpler netting analysis to determine the maximum burst pressure, see ref. [6.28], where it is assumed that only the fibres carry load.

Alternatively the designer may require a vessel which, because of attachments to other components, has the minimum axial or radial expansion when pressurised. This can be determined from a design

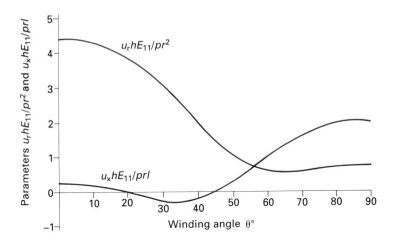

6.22 Design chart for radial u_r and axial u_x displacements in a GRP cylindrical pressure vessel (materials data as for Fig. 6.21).

chart for the displacement u_x, u_r. In equations [6.89] the elastic constants E_{xx}, E_{yy}, ν_{xy} are all functions of the winding angle θ. These functions are readily determined from CLT and were computed with LAMICALC for an angle-ply GRP laminate. Using the computed results in equation [6.89] the design chart Fig. 6.22 was derived.

Figure 6.22 shows that for $\theta < 45°$ u_x is always small and $u_x = 0$ at $\theta = 20°$ and $44°$, showing that at these winding angles there is no axial extension in a GRP vessel. The radial extention u_r is seen to be lowest for $\theta > 60°$.

6.5.2.2 Further references
For further information on the design of composite pressure vessels, see refs [6.5] and [6.28]. In ref. [6.5] the extension of the analysis given here to thick-walled vessels is also given, together with a discussion of the PC program MIC-MAC which is suitable for the analysis of thin-walled vessels. The PC program CYLAN[6.18] is also valuable for the analysis of composite cylinders under pressure, axial and torsional loads and for analysing certain buckling and vibration effects. For more general information on the design of vessels and pipes, with particular attention to practical details such as joints cut-outs, etc., the reader is referred to codes of practice and standards such as those for GRP vessels in BS 4994[6.29] and GRP pipes and fittings in BS6464.[6.30]

6.6 Concluding remarks

This chapter has attempted to present a series of design procedures for composite structural elements. Although the procedures are simple to use, they follow from a more detailed study of anisotropic structures, often backed up by experimental test programmes. They are based on idealised geometry and are suitable for preliminary design analysis, for determining fibre lay-ups, wall thicknesses, etc., as in the iterative design loop shown in Fig. 6.1. For many structures such simplified analyses will be sufficient, especially when carried out in conjunction with a similar analysis of joints and load introduction as discussed in Chapter 8. In other cases the preliminary design analysis will need to be followed by more detailed design analysis which is now conveniently carried out using finite element analysis (FEA). In these cases the preliminary design provides valuable input data for the FEA, and may considerably reduce its cost by providing information on the best choice of materials so that parameters such as fibre angles do not need to be investigated iteratively during the FEA.

Conventional FEA software may be used, in which case the composite's laminate analysis is uncoupled from the FEA, or with the newer composites, FEA software laminate analysis and FEA are integrated. The conventional software may be used for analysing certain simpler composites structures based on orthotropic plate and shell elements. In this case a laminate analysis may be required using CLT software such as MIC-MAC[6.5] or LAMICALC[6.27] to determine the effective elastic constants of the orthotropic plate element. With these as input material data, the FEA then computes structural deflections and element stresses. Failure analysis may be based simply on the study of maximum stresses for comparison with uniaxial laminate strength values. A better, but time-consuming, approach is to identify the critical elements and then use the element stresses as input data for CLT in order to determine strength reserve factors.

Conventional software is valid, or will give a reasonable approximation, for composite materials which contain a single type of reinforcement, with all plies parallel, e.g. mat and fabric reinforced plastics or UD laminates. In these cases the implicit assumption of a homogeneous orthotropic material is a good approximation, so that the analysis is valid and the post-processing of element stresses is also meaningful. However, for laminated composites with different ply angles, or for hybrid composites and sandwich materials, the use of a standard shell element is not valid, since for these materials the in-plane and flexural stiffnesses are different and the material may also exhibit coupling effects such as bending–torsion or bending–extension coupling. The FEA must then be based on orthotropic thick shell elements which contain independent stiffnesses for membrane and bending stresses. Such elements are incorporated into the newer composites FEA software, such as P/COMPOSITE,[6.31] ANSYS,[6.32] PERMAS-LA,[6.33] COMPOSIC[6.34] and SAMCEF.[6.35]

With this specialised composites FEA software the laminate analysis is coupled with the FEA and the material's input data are the lay-up and ply properties. A preprocessor in the software uses CLT, or an extended CLT with shear effects as in ref. [6.33], to compute the element stiffness properties in the $[A]$, $[B]$ and $[D]$ stiffness matrices. Nodal displacements and element stress resultants are then computed in the FEA. In postprocessing the stress resultants are used within CLT to compute ply stresses and ply strength reserve factors based on an appropriate composite's multiaxial failure criteria. The latter may be displayed graphically so that the designer can determine the critical stress regions in the structure and see which ply fails first in these regions.

This new, integrated composites software is a powerful tool for the design analysis of composite structures. However, to use it effectively requires considerable experience of both composite materials mechanics and FEA. Even with these advanced design tools, there is still an important role for analysis based on simplified methods such as those described here for the preliminary design analysis of the structure.

6.7 References

[6.1] J.R. Vinson and R.L. Sierakowski, *The Behaviour of Structures Composed of Composite Materials*, Martinus Nijhoff, 1987, Dordrecht/Boston/Lancaster.

[6.2] J. Wiedemann, *Leichtbau*, Vol I Element, Springer-Verlag, Berlin, Heidelberg, 1986. Vol II Konstruktion, Springer-Verlag, Berlin, Heidelberg, 1989.

[6.3] A.F. Johnson and A. Marchant, 'Design and analysis of composite structures', Ch. 3 in *Polymers and Polymer Composites in Construction* (ed. L. Hollaway), London, Thomas Telford, 1990.

[6.4] W.C. Young, *Roark's Formulas for Stress and Strain*, 6th edition, New York, McGraw-Hill, 1989.

[6.5] S.W. Tsai, *Composites Design*, 4th edition, with MIC-MAC PC program, Dayton, Ohio 45419, Think Composites, 1988.

[6.6] H.G. Allen and P.S. Bulson, *Background to Buckling*, London, New York, Tokyo, McGraw-Hill, 1980.

[6.7] J.M. Whitney, *Structural Analysis of Laminated Anisotropic plates*, with LAMPCAL PC program, Lancaster PA 17604, Technomic, 1987.

[6.8] ESDU 80023, *Buckling of Rectangular Specially Orthotropic Plates*, and PC programs M2018/2021, London, ESDU International, 1980.

[6.9] W.J. Stroud and N. Agranoff, *Minimum Mass Design of Composite Panels Under Combined Loads: Design Procedure Based on Simplified Buckling Equations*, NASA TN D-8257, 1976.

[6.10] A.F. Johnson and A. Woolf, 'Deflection and stress analysis of orthotropic plates in flexure', *Computers and Structures* 1984, 18/5, 911–919.

[6.11] A.F. Johnson and G.D. Sims, 'Simplified design procedures for composites plates under flexural loading' in *Proc. 2nd International Conference on Composite Structures* (ed.) I. Marshall, London, New York, Applied Science, 1983.

[6.12] A.F. Johnson and G.D. Sims, *Design Procedures for Plastics Panels*, with PC program PANDA, London, National Physical Laboratory, 1987.

[6.13] ESDU 83036, *Natural Frequencies of Laminated Flat Plates*, and PC Program M3007, London, ESDU International, 1989.

[6.14] ESDU 2034, *Stress Concentrations around Circular Holes in Composite Panels*, and PC program M2034, London, ESDU International, 1989.

[6.15] A.F. Johnson, 'Design of RP Cylinders Under Buckling Loads', in *Proc. 40th Reinforced Plastics/Composites Institute*, Annual Conference Paper 10-E SPI, New York, 1985.

[6.16] A.F. Johnson, *Simplified Buckling Analysis for R P Beams and Columns*, ECCM-1, Bordeaux, 1985.

[6.17] C.C. Chamis and P.L.N. Murthy, 'Design procedures for fibre composite box beams', *J. Reinforced Plastics and Composites* 1989, **8** 370–97.

[6.18] CYLAN, Cylinder Analysis Program, Lancaster PA 17604, Technomic, 1987.

[6.19] A.F. Johnson and G.D. Sims, 'Performance analysis of pultruded composites sections', *Composites Polymers*, 1989, **2**(2), 89–112.

[6.20] G.D. Sims, A.F. Johnson and R.D. Hill, 'Mechanical and structural properties of a GRP pultruded section', *Composite Structures*, 1987, **8** 173–87.

[6.21] A.F. Johnson and G.D. Sims, 'Mechanical properties and design of sandwich materials', *Composites*, 1986 **17**(4) 321–328.

[6.22] H.G. Allen, *Analysis and Design of Structural Sandwich Panels*, Oxford, Pergamon, 1969.

[6.23] US Department of Defence, *Structural Sandwich Composites*, MIL-HDBK 23A Washington, 1968.

[6.24] K.F. Morgan, 'Design to the limit: optimising core and skin properties', in FIT 3rd International Conference 'Marine Applications of Composite Materials', Melbourne, Florida, 1990.

[6.25] ESDU 88015, *Wrinkling of Sandwich Panels with Composite Faceplates*, and PC program M2036, London, ESDU International, 1988.

[6.26] ESDU 85037, *Natural Frequencies of Sandwich Panels with Laminated Faceplates*, and PC Program M3015, London, ESDU International, 1988.

[6.27] LAMICALC Laminate Analysis Program, Composites Software, Drosselweg 7, D-7046, Gaufelden 2, 1992.

[6.28] G.C. Eckold, 'A design method for filament wound GRP vessels and pipework', *Composites* 1985 **16**(1), 41–47.

[6.29] British Standards Institution, *Design and Construction of Vessels and Tanks in Reinforced Plastics*, BS 4994, London, BSI, 1981.

[6.30] British Standards Institution, *Specification for Reinforced Plastic Pipes, Fittings and Joints for Process Plants*, BS 6464, London, BSI, 1984.

[6.31] PATRAN P/COMPOSITE, PDA Engineering, Cosa Mesa, CA 92626.

[6.32] ANSYS, Swanson Analysis Systems Inc, Houston PA 15342-0065.

[6.33] PERMAS-LA, Intes GmbH, D-7000 Stuttgart 80.

[6.34] COMPOSIC, CSI, F-60471 Compiegne.

[6.35] SAMCEF, CISI Ingeneric SA, F-94528 Rungis.

7 Limit state design method

7.1 Introduction

Limit state design principles have become internationally recognised as the basis of codes of practice for a wide range of structural applications and materials. However, their application to design using FRP materials has been very limited, mainly because of absence of design codes using such principles but also because of the lack of familiarity with these principles among designers traditionally working with FRP materials. Nevertheless, future codes of practice on the structural application of FRP materials are likely to be based on limit state principles, particularly for codes developed internationally through such as the European Standardisation Committee (CEN). Indeed the European Community directives for European standards harmonisation recommend that all such harmonised standards be based on limit state principles for structural design. It is therefore advisable for FRP materials designers to become familiar with these principles as they apply to their materials, not only so that they can work with these codes but also so that they are able to demonstrate that products for which no design codes exist (as will often be the case) have equivalent levels of safety and reliability incorporated into their design.

The aim of this chapter is therefore to provide an introduction to the subject for those familiar with FRP materials but not with limit state design (or vice versa). More detailed descriptions can be found in refs [7.1]–[7.3].

7.2 General principles of limit state design

7.2.1 Aims of limit state design

The document forming the basis of the development internationally of limit state design codes for various materials[7.1] defines the aim of design as:

the achievement of acceptable probabilities that the structure being designed will not become unfit for the use for which it is required during some reference period and having regard to its intended life. Thus, all structures or structural elements should be designed to sustain, with an appropriate degree of safety, all loads and deformations liable to occur during construction and proper use, to perform adequately in normal use and to have adequate durability during the life of the structure.

It is useful to replace the term 'safety' with the quantitative term 'reliability', which is defined as the complement to the probability of an adverse event, for example, failure. One of the advantages of limit state design is that the method is readily adaptable to give the required reliability for a particular type of structure or product, and that correlation can therefore be achieved between design standards using different materials. A further advantage is that it provides a logical framework within which the uncertainties in test data, loading, stress analysis, etc., can be quantified and understood in a consistent manner between different materials, whilst still recognising the varying characteristics of those materials. This is especially significant in the context of FRP materials which are themselves highly variable in characteristics.

7.2.2 Levels of sophistication

Three levels of sophistication are recognised in checking the degree of safety in structures at a given limit stage:

> *Level 3*: an 'exact' probabilistic design method involving integration of the representation of the various uncertainties to derive optimum failure probabilities.
>
> *Level 2*: a probabilistic design method in which the basic variables are represented by their known or postulated distributions and a defined reliability level is accepted. The distributions are assumed to be uncorrelated.
>
> *Level 1*: a semi-probabilistic design method in which appropriate levels of reliability are provided by the specification of a number of partial safety factors related to nominal values of the basic variables.

Level 3 concepts are used primarily in research. Design codes generally adopt a Level 1 approach, although Level 2 concepts will normally have been used in the design of the code itself. Level 2 methods may be used directly in design if sufficient statistical data for the basic variables are available (or can be obtained) and the required level of reliability justifies the cost. However, it is envisaged that for the majority of FRP applications Level 1 design methods will be the most appropriate, once appropriate codes of practice have been developed, and it is this method which is described in more detail here.

7.2.3 Limit states

The performance of all or part of an FRP structure or structural component can be described with reference to a limited set of limit states beyond which the design requirements are no longer satisfied. Limit states can be placed into three categories:

1   The ultimate limit states (ULS), corresponding to the maximum load-carrying capacity.
2   The serviceability limit states (SLS), relating to the criteria governing normal use.
3   The conditional limit states, corresponding to an infrequent major random event, e.g. fire. Conditional limit states are frequently placed in one of the other categories (ULS or SLS).

Examples of the ultimate limit states include loss of stability (buckling) and rupture of critical sections due to the material strength being exceeded (strength may be reduced by environmental effects or by repeated loading). Serviceability limit states may include excessive deformation, excessive vibration and local damage which reduces durability or affects efficiency or appearance.

7.2.4 Design

All the relevant limit states need to be considered in the design of an FRP structure, although it is generally acceptable to design on the basis of the limit state judged to be critical and then to check that the remaining limit states are not reached.

The consideration of each limit state requires the setting up of a calculation model incorporating the appropriate basic variables allowing for the influence of direct and indirect loads, the response of the structure and the behaviour of the material. The uncertainties associated with the variables and with the calculation model itself are allowed for either by a method of partial coefficients (Level 1 approach) or by a probabilistic method (Level 2). The basic variables are considered to be independent, random variables.

7.2.5 The method of partial coefficients (Level 1)

In the method of partial coefficients the variability of these basic variables are taken into account by the selection of characteristic values for each of them. The uncertainties associated with the characteristic values and the calculation models are taken into account by the application of partial coefficients to the variables or their effects. Where the necessary data are available, the characteristic value for a variable is normally based on a statistical interpretation of that data. Thus for a varying load the characteristic value is defined as that value that has a prescribed probability of not being exceeded within the reference period. When characteristic values for loads cannot be established from statistical data, they could be estimated on the basis of available information, and, possibly, forecasts of future developments.

The variation of strength and other properties of the materials are treated by defining characteristic properties which would be related to standard test specimens and procedures. Standardised test procedures would be vital for the successful application of the method. The characteristic value of a material property is normally defined as that value that has a prescribed probability of not being reached in a hypothetical unlimited test series.

Two types of partial coefficient are used, one type for the loads or load effects and the other for the strength of materials or elements. The partial coefficients vary depending on the load type, the material, the type of FRP structure or component, the importance of the application, and the safety and economic consequences of the limit state under consideration being exceeded. When partial coefficients are applied to a characteristic value, the resulting value is termed a design value. The design value for a particular variable might vary for different limit states (whereas the characteristic value is normally constant). Thus, for a particular limit state, the design value of the different types of variable would be given by:

$$F_d = \gamma_f F_k$$

$$f_d = f_k / \gamma_m$$

where F_d is the design value of a load, F_k is the characteristic value of the load, γ_f is the partial coefficient for the load, f_d is the design material property (e.g. strength), f_k is the characteristic property and γ_m is the partial material coefficient for the property.

The partial coefficients can be further subdivided to enable rational and consistent values for the individual coefficients to be defined. Thus the partial coefficient γ_f can be considered as a function of three coefficients, γ_{f1}, γ_{f2} and γ_{f3}, where: γ_{f1} takes account of the possible unfavourable deviation of the loads from the characteristic values; γ_{f2} takes account of the reduced probability that combinations of loads will simultaneously reach their characteristic value; and γ_{f3} takes account of possible inaccurate assessment of the load effects.

The partial material coefficient γ_m can also be broken down, and can be considered as a function of two coefficients γ_{m1} and γ_{m2}, where: γ_{m1} takes account of unfavourable deviations of the material properties from the specified characteristic values, and the possible differences between the material property in the structure or element and that derived from test specimens; and γ_{m2} takes account of possible local weaknesses arising from the manufacturing process, and unfavourable geometric deviations resulting from manufacturing tolerances. A greater number of subdivisions can be chosen if necessary.

A further coefficient, γ_n can be introduced to adjust either γ_f or γ_m. γ_n can be considered as a function of two coefficients, γ_{n1} and γ_{n2}, where: γ_{n1} takes account of the nature of the failure associated with the limit state, e.g. whether ductile or brittle and whether collapse can occur without warning; and γ_{n2} takes account of the consequences of failure, from economic considerations as well as those of human safety. However, γ_n is normally used simply as a modifying factor to the other factors rather than explicitly as a separate factor.

Fabrication tolerances are recognised as a problem to both mould-ers and designers. In certain cases where the variability of a geometric parameter could have a significant effect on the strength of an element, for example the thickness of a thin composite skin, it would be appro-priate to apply a coefficient explicitly rather than by inclusion in γ_m. The design values a_d of geometrical parameters would be obtained from the characteristic values a_k (normally the nominal specified values) and an additive element Δ_a, since additive elements would generally be more suitable than factors for geometric parameters, i.e.

$$a_d = a_k + \Delta_a$$

where Δ_a takes account of the importance of variations in a, and the given tolerance limits for a.

7.2.6 Choice of values for the partial coefficients

The values of the partial coefficients depend on the limit state under consideration and should be based on statistical data when available. Where such data are not available the coefficients are derived by cali-bration with pre-existing practice or judgement.

Where a limit state design Code of Practice covers the material and application being considered, this would of course normally specify the partial coefficients to be used for each limit state (the above process having been carried out by the authors of the Code).

7.2.7 Summary of the partial coefficient method

The object of the partial coefficient method outlined above for FRP structures is to achieve, by means of suitably derived partial coeffi-

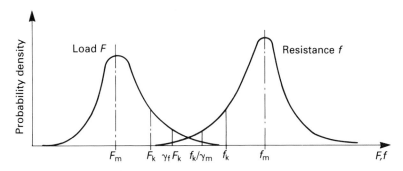

7.1 Load and resistance distributions.

cients applied to the characteristic values of the relevant basic variables, acceptable probabilities that for each limit state the resistance of the structure or element would exceed the load effects. This is illustrated in simple form in Fig. 7.1, where a typical resistance function, f, and a typical load function, F, are plotted.

7.3 Characteristics of FRP materials

7.3.1 General

The great variety of polymers, additives and modifiers means that there are no universal methods describing the behaviour of all polymer materials. Nevertheless, polymers do exhibit many similar characteristics, with differences only in magnitude, and for the most part it is these characteristics that distinguish the structural behaviour of polymers from those of conventional materials. The more pronounced viscoelastic behaviour and influence of temperature and environmental factors are examples of this. The many types of fibre reinforcement would appear to introduce further complexity. However, in some respects the behaviour of FRP is less complex than unreinforced and short-fibre plastics, because typically both the fibre reinforcement and the polymer matrix are more elastic (i.e. less viscous) than thermoplastics, resulting in lesser dependence of properties on temperature and duration of loading. In other respects, of course, the behaviour is more complex, arising from the wide disparity in properties of the constituents and the possible variations in the arrangement of the fibres. All of these factors need to be embraced by the limit state design method.

In the following sections the main characteristics of FRP materials are described, drawing particular attention to where these differ from conventional structural materials e.g. concrete, metals, masonry, wood) so that those aspects that need to be covered by the design method are summarised. In this context the definition of FRPs excludes randomly oriented isotropic short-FRP materials.

7.3.2 Structural characteristics of FRP materials

7.3.2.1 Fibres
Fibre reinforcement not only comes in a variety of materials with different strengths and stiffnesses, but also in a variety of forms, e.g. mats, straight rovings, woven fabrics. In some forms the fibres are grossly kinked to conform to a weave pattern and this can reduce the strength of the composite material. Unlike most conventional materials, the strength and stiffness of the material can be varied by adjusting the fibre content.

7.3.2.2 Matrices
The wide variety of polymers with different characteristics is further complicated by the addition of fillers and plasticisers which can significantly alter the composite properties. This can result in published properties being of little value because the exact composition is not stated or cannot be reproduced. Some matrices may exhibit poor bonding with the fibre reinforcement and are thus unable to develop the full strength capacity of the fibres.

7.3.2.3 Processes
The characteristics of materials with identical constituents can be influenced by the process used in their production. Factors such as unplanned fibre orientation, insufficient compaction, inadequate curing, inaccurate fibre placement and lap joints in preformed reinforcement can all adversely affect the material properties.

7.3.2.4 Short term stress–strain relationships
Provided fibres are orientated in the stress direction most FRP materials exhibit mainly linear elastic behaviour to failure. Characteristic stress-strain curves are shown in Fig. 7.2. Type I is the most typical for materials

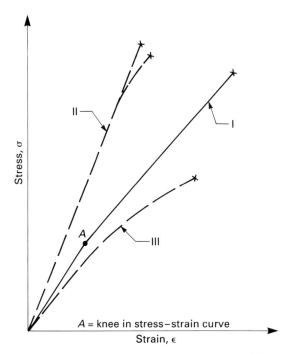

7.2 Characteristic stress–strain curves for reinforced plastics.

with fibres aligned in more than one direction, with the 'knee' indicating the onset of first damage, i.e. crazing or microcracking at the transverse fibres. Type II is typical for unidirectional laminates stressed parallel to the fibres and Type III is typical for cases where the influence of the matrix is significant, i.e. random in-plane reinforced laminates or aligned fibre laminates stressed off the principal fibre axes.

Unlike most structural materials, FRP materials do not exhibit yielding (except for aramid fibre composites in compression) and therefore are not ductile. However, they do not generally display the other characteristics of classic brittle materials, i.e. low tensile strength and sensitivity to impact and stress concentrations. This is because the fibres act as crack inhibitors, although where there are no fibres to limit the spread of cracks, e.g. transverse to a unidirectionally reinforced laminate or in interlaminar layers, the material behaves in a brittle manner.

7.3.2.5 Directional properties
FRP materials are neither homogeneous nor, generally, isotropic. Depending on the fibre alignment, an FRP laminate can exhibit in-plane elastic properties which are planar-isotropic, orthotropic or aniso-tropic. In aligned fibre laminates properties can alter significantly when the direction of applied stress departs from a principal fibre direction. Unbalanced and asymmetric multi-ply laminates show even more complex behaviour.

7.3.2.6 Long term loading
Under sustained stress, the strain in FRP continues to increase and the magnitude of stress needed to produce rupture diminishes with time. These phenomena are referred to as creep and creep rupture respectively. The behaviour is complex and is influenced by temperature and environment. Although the effect in FRP is less pronounced than with unreinforced polymers, it is generally more significant than with other structural materials.

7.3.2.7. Fatigue
As with static loading, fatigue behaviour under repeated or cyclic loading is highly directional. Unlike most conventional structural materials, some FRP materials, notably glass reinforced materials, do not have a fatigue limit. High frequency stress variations can generate internal heat in the material.

7.3.2.8 Environmental factors
The interaction of time, temperature, stress and environment can have a profound effect on the structural performance of FRP materials, to a far greater extent than with any other major structural material except

possibly wood. Even normally occurring agents such as water (or water vapour) and ultraviolet rays from the sun can seriously affect FRP materials.

7.3.2.9 Characterisation of FRP materials

Other structural materials, such as wood and concrete, display widely varying properties and characteristics. However, design in these materials is facilitated by classification of the materials, related to established specifications in the case of concrete, and by generally accepted simplifications of the overall behaviour. No such characterisation exists in FRP materials.

7.4 Application of partial coefficient method

7.4.1 General

When it comes to applying limit state methodology to an actual FRP design exercise, the ideal situation would be one where there existed an accepted national or international Code of Practice, based on limit state philosophy and covering the intended materials and application. Such a code would normally set out the limit states to be considered and specify characteristic values and partial coefficient values to be used. However, few such codes exist at present and it will be found necessary in most cases to develop the design criteria from general principles (as must be done for other design methodologies where no code exists).

The broad range of applications of FRP materials makes it inadvisable to attempt to develop a single set of design criteria to treat all possible applications. A similar approach is adopted for other materials, e.g. there are different design criteria for steel buildings and steel offshore structures. Design methods and national codes will tend to be application-orientated and the following paragraphs endeavour to set out the general principles on which limit state design criteria for FRP structures could be based.

It is not proposed to discuss criteria for loads since these are essentially independent of the structural material.

7.4.2 Limit states

For the design of any structure or component it is necessary to select all the relevant limit states for consideration. The actual limit states selected obviously depend on the nature of the FRP structure and its intended use, and the following is intended only as an indication of those limit states that have to be considered in view of the particular characteristics of FRP described previously.

7.4.2.2 Serviceability limit states (SLS)

1 Deflection/deformation: this is to be a critical limit state in many applications because of the generally lower magnitude and directional nature of FRP stiffness compared with conventional structural materials such as steel and concrete.
2 Residual deformation: the viscoelastic behaviour of some FRP materials may result in excessive permanent deformation under sustained loading.
3 First damage: crazing or microcracking of the material may constitute a limit state for various reasons, e.g. it accelerates environmental degradation or causes leakage from a fluid container.
4 Buckling or wrinkling (local): this may be a serviceability or ultimate limit state depending on the nature of the application and the post-buckling behaviour of the material.
5 Vibration.
6 Environmental damage, i.e. effect on performance (reduced properties) or appearance without causing failure.
7 Impact damage: as (6).

7.4.2.3 Ultimate limit states (ULS)

1 Collapse, i.e. loss of equilibrium of whole or part of structure.
2 Rupture of section: causing failure of structure or element of structure. This may take several forms, e.g. tension, shear, flexure or interlaminar shear, and may require definition of failure criteria for the material. Rupture may be influenced by the interaction of time, temperature, stress and environment.
3 Buckling, i.e. loss of stability. May be critical owing to lower stiffness and directional nature of properties.
4 Fatigue.
5 Resonance.
6 Fire resistance.

7.4.3 Methods of analysis (ULS)

Methods of analysis at ultimate loads have to take account of the lack of ductility in most FRP materials, in contrast to most conventional structural materials. This has two main effects:

1 Non-linear structural analysis involving redistribution of stresses in redundant structures is not valid.
2 Internal redistribution of the stresses in a cross-section cannot take place. This means that load effects such as temperature gradients and shear lag have to be considered at the ultimate limit state (unlike ductile materials where these effects need only be considered at the serviceability limit state).

However, the almost linear stress–strain curves to failure of most FRP materials means that elastic methods of analysis are adequate for most applications.

7.4.4 Material parameters

7.4.4.1 Characteristic material properties
The required characteristic material properties for FRP materials differ in many respects from those required for most conventional structural materials. Firstly, standard grades of material are not available. Steel, for example, is produced under stringent quality control to a nominal guaranteed yield stress, which is then used as the characteristic strength in limit state design, and the elastic modulus and the Poisson ratio vary little, whatever the composition. The material is homogeneous and isotropic, so no other elastic constants are required and the properties vary little under normal temperature ranges and environmental conditions.

By contrast, standard data on FRP composites is generally either not available or cannot be relied upon except for preliminary design. Up to 21 elastic constants may be required for a rigorous analysis of a complex laminate, stiffness as well as strength varies with the composition, and the properties vary with temperature and production process and may degrade with time. Material design data are therefore normally derived from the testing of specimens. For maximum reliability the specimens and test conditions should represent as closely as possible the materials and conditions of use of the final product.

Derivation of characteristic design properties from test data is a specialist subject and is dealt with elsewhere, for example refs [6.4] and [6.5]. In the construction industry characteristic material properties are commonly taken as the lower 5% fractile, but other values may be appropriate for different applications. The quoted characteristic values, which for FRP materials will include stiffnesses, strengths and possibly first damage as well as ultimate strengths, should be associated with the governing limitations, e.g. constituent materials, production process, limits of temperature and environmental conditions. This implies, of course, that separate characteristic values should be used when any of the conditions is changed, e.g. a different temperature range. An alternative method would be to use a single characteristic value for each property, for a given set of conditions, and to use factors to derive representative values for other conditions. These would not be true partial coefficients, however, since they do not represent uncertainties.

7.4.4.2 Partial material coefficients
The partial material coefficient γ_m allows for uncertainties in the assumed properties of the material in the final structure. Ideally it

should be derived from statistical data but this requires specialist techniques[6.4] and sufficient data are rarely available (since specimen testing does not deal with all the uncertainties). The coefficient is normally derived by a combination of statistical methods and calibration with pre-existing practice, or simply by judgement where there is not relevant pre-existing practice.

Most existing limit state design codes do not subdivide the γ_m factor, although different values are frequently given for different limit states, indicating that the derivation involved subdivision. FRP material properties used in design tend to have uncertainties not associated with more conventional materials. One such uncertainty is in the method of deriving properties. For the properties of multi-ply laminates, for example, there are three different methods of deriving the laminate characteristics, with reducing level of uncertainty:

1 Properties of constituent materials (fibre, matrix) derived from test specimen data; properties of individual plies derived from theory; properties of laminate derived from theory.
2 Properties of individual plies derived from test specimen data; properties of laminate derived from theory.
3 Properties of laminate derived from test specimen data (not always practical).

Clearly different partial coefficient values should be applied to the characteristic value depending on which derivation method is used, and this may justify a further subdivision of the partial material coefficient. Different production processes may also justify different partial coefficient values (in addition to different characteristic values) in view of the different level of uncertainty associated with each process. Materials produced by automated processes generally demonstrate a lower spread of test results and are produced to smaller dimensional tolerances than manual processes, and this can be reflected in lower partial coefficient values and a resulting economy in design.

7.4.4.3 Brittle failure modes

Higher values of the partial material factor are normally applied when the failure mode is brittle. Thus tensile failure of a unidirectional composite, failure by interlaminar shear or tension, and a failure of certain bonded connections are examples of limit states where higher values would apply.

7.4.5 Buckling instability

Two distinct methods of dealing with buckling instability can be found in existing limit state design codes. In those dealing with concrete structures, slender compression members must be designed for an additional eccentricity of load, the magnitude depending on the slenderness ratio of the member. In those dealing with metal structures, local plate buckling and overall buckling instability are dealt with by strength reduction factors dependent on slenderness ratios, and this would appear to be the more appropriate method for FRP structures. Additionally, for FRP structures and plates in compression, uncertainties in the stiffness parameters need to be allowed for by the partial material coefficient.

7.4.6 Fatigue

Again, two distinct methods have been formulated in other limit state codes. In one draft code it is proposed that the calculated fatigue life should exceed the design life factored by a partial coefficient. However, a more rational approach is to use characteristic S–N curves (i.e. curves joining lower fractile points rather than the mean values) to calculate fatigue life, which is then required to exceed the design life.

For FRP materials, fatigue behaviour is influenced by temperature, environment and direction of stress (for non-isotropic laminates), and characteristic S–N curves should be associated with governing limitations as suggested for other material properties, Section 7.4.1.

7.4.7 Summary

The preceding paragraphs demonstrate that limit state design philosophy can be applied to the design of FRP structures. Those difficulties which may occur in representing the complex behaviour of FRP would occur whichever design philosophy is used, and the benefits of using a logical and rational approach consistent with other structural materials are considerable.

7.5 Example

The following fictitious example has been selected to give a simple illustration of the application of limit state design principles to an FRP component. The characteristic and partial coefficient values have not been derived by statistical or any other methods, and should not be regarded as proposed values.

An electronic scoreboard is to be suspended from the roof of a covered stadium by a number of hangar rods. The design criteria for each rod are listed below:

Length: 8.0 metres
Diameter: to be determined
Loading: 110 kN dead load;
            15 kN live load (occasional maintenance);
            33 kN (30% DL) accidental overload due to failure
            of one other hangar.

Deflection: 15 mm maximum during design life (due to dead load).

Design life: 10 years.

The selected material is graphite–epoxy composite with protection against environmental degradation. The characteristic short term material properties (from tests) are:

Longitudinal modulus, E = 180 GPa
Ultimate tensile strength, σ_μ = 1500 MPa

Sustained loading tests indicate that rupture occurs at 80% of ultimate tensile strength after 10^5 hours (11.4 years).

7.5.1 Limit states

The following limit states will be considered:

Ultimate (ULS):   1.   Creep rupture
                        2.   Accidental collapse

Serviceability (SLS):   3.   Deflection
                        4.   First damage

The rod will be designed for the creep rupture limit state and then checked against the remaining limit states.

7.5.2 Partial coeffients

Partial coefficients are given in Table 7.1.

7.5.3 Material property reduction factors

Creep: graphite–epoxy composites exhibit little creep deformation, but a stiffness reduction factor of 1.1 will be used to allow for what may occur within the design life.

Table 7.1 Partial coefficients

		ULS	SLS
Partial load coefficient, γ_f:	Dead load	1.5	1.2
	Live load	2.0	1.5
	Accidental	1.2	—
Partial material coefficient, γ_m:	Strength	3.0	—
	Stiffness	—	1.1

Creep rupture: strength reduction factor of 1.25 will be used to allow for the reduction in strength due to sustained loading over the reference period.

7.5.4 Design

7.5.4.1 Creep rupture
The reduced characteristic material strength for sustained loading, f_k:

$$f_k = 1500/1.25$$
$$= 1200 \text{ MPa}$$

Design material strength (ULS), f_d:

$$f_d = f_k/\gamma_m$$
$$= 1200/3.0$$
$$= 400 \text{ MPa}$$

Design load (ULS):

$$F_d = \gamma_f F_k$$
$$= 1.5*110+2.0*15$$
$$= 195 \text{ kN}$$

The required cross-section area A of the rod is given by:

$$195\,000/400 = 487.5 \text{ mm}^2$$

therefore required diameter of rod, $D = 25$ mm.

7.5.4.2 Accidental collapse
The design load for this limit state includes dead load only plus accidental load, therefore design load (ULS):

$$F_d = \gamma_f F_k$$
$$= 1.2\,(110+33)$$
$$= 171.6 \text{ kN}$$

This is less critical than creep rupture.

7.5.4.3 Deflection
For ten year design life the material stiffness reduction factor is 1.1, therefore the characteristic viscoelastic modulus is given by:

$$E_{vk} = E/1.1$$
$$= 180/1.1$$
$$= 163.6 \text{ GPa}$$

Design viscoelastic modulus (SLS):

$$E_{vd} = E_{vk}/\gamma_m$$
$$= 163.6/1.1$$
$$= 148.8 \text{ GPa}$$

Design load (SLS – dead load only):

$$F_d = \gamma_f F_k$$
$$= 1.2*110$$
$$= 132 \text{ kN}$$

therefore long term deflection is given by:

$$d = F_d L/(E_{vd}A)$$
$$= 132\ 000*8000/(148\ 800*490)$$
$$= 14.5 \text{ mm}$$

i.e. less than the permitted value (15 mm).

7.5.4.4 First damage

The criterion adopted here is a maximum elastic tensile strain of 0.2%. Design material modulus (SLS):

$$E_d = E/\gamma_m$$
$$= 180/1.1$$
$$= 163.6 \text{ GPa}$$

Design loading (SLS):

$$F_d = \gamma_f F_k$$
$$= 1.2*110+1.5*15$$
$$= 154.5 \text{ kN}$$

therefore the strain is given by:

$$e = F_d/(E_d A)$$
$$= 154\ 500/(163\ 600*490)$$
$$= 0.19\%$$

i.e. less than the permitted value (0.2%).

7.5.6 Conclusion

A 25 mm diameter rod with the specified properties satisfies all the critical limit states and is therefore acceptable.

7.6 References

[7.1] *Common Unified Rules for Different Types of Construction and Materials*, Vol. 1, International System of Unified Standard Codes of Practice for Structures, CEB-FIP International Recommendations, 3rd edition, 1978.

[7.2] Joint Committee on Structural Safety (JCSS), *General Principles of Reliability for Structural Design*, International Association for Bridge and Structural Engineering, Report 35, Part II, 1981.

[7.3] A.L.L. Baker, 'Criteria of structural design,' *Proceedings of the Institution of Civil Engineers* Vol 56 Part 1 Nov 1974 407–427.

[7.4] *Rationalisation of Safety and Serviceability Factors in Structural Codes*, CIRIA Report 63, Construction Industry Research and Information Association, July 1977.

[7.5] R.L. King, 'Statistical methods for determining design allowable properties for advanced composite materials', British Plastics Federation, Reinforced Plastics Congress 1986.

Part IV

Joints

8 Bonded and mechanically fastened joints

8.1 Introduction

Joints may be needed:

1 Because of limitations in material size.
2 For convenience in manufacturing or transportation.
3 To provide access.
4 To link sub-assemblies.

Wherever possible a designer should avoid using joints.

The main methods used for metals, adhesive bonding and mechanical fastening are also applicable to composites. Welding may be a possibility for thermoplastic composites, but the method is not well developed for load-carrying joints.

When selecting a method the issues shown in Table 8.1 should be considered. It is always possible to design a bonded joint that is stronger than the composite being joined. It is very difficult to design a mechanically fastened joint to have a strength greater than 50% of that of the composite; to obtain acceptable efficiency, local reinforcement may therefore be needed.

Galvanic corrosion of certain meals (aluminium and alloys, low alloy steels, martensitic steels, copper, brass, cadmium (plating)) can occur when using carbon fibre-reinforced polymers (CFRP). Use of a sealant or corrosion protection barrier is recommended at the interfacing surface of composite-to-metal joints, or when using fasteners, in these materials.

Table 8.1 Issues to be considered when selecting a method for fastening

Issue	Adhesively bonded	Mechanically bonded
Is joint disassembly required?	Not possible	Possible
Will surface preparation be needed?	Yes	No
Will NDI be needed?	Yes	No
Will there be high stress concentration?	No	Yes
Will there be a weight penalty?	No	Yes
Will the joint be sensitive to environment?	Yes	No
Cost?	Low	High
Will accurate assembly be needed?	Possibly	No

Joints must be considered at the start of the design process, as should the need for repairability.

8.2 Joint types

The basic joint types considered in this chapter are shown in Fig. 8.1 and they can all be used with either bonded or mechanical fastening. However, special requirements exist when using mechanical fasteners with tapered thickness plates. Strap joints can be considered as variants of single or double laps. The simplest joint, i.e. unsupported single lap, is the weakest; the strongest configuration being the most complicated, i.e. the stepped joint.

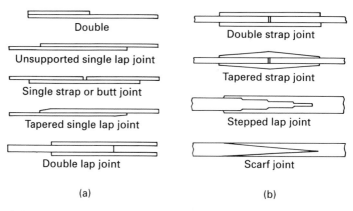

8.1 Basic joint types (after ref. [8.4]).

8.3 Mechanism of load transfer

The purpose of a joint is to transfer load between the two items being joined. As a result of this load transfer there will be a stress variation in the components in the joint region, as well as stresses in the joining medium (fasteners or adhesive).

The mechanism is illustrated for a single cover butt joint in Fig. 8.2. The load is transferred from plate A to plate B via the cover plate C. The plate stresses in the direction of the load will be zero at the points marked x, the free ends, and maximum at the points marked y. The variation of strain in the plates along the joint means that the adhesive or the fasteners will be loaded in shear, as shown in Fig. 8.2 (b) and (c). Because the lines of action for the resultant load in the plates and cover plate are not coincident, the joint will bend under load, as shown in Fig. 8.2(d). The joint will tend to peel apart at points x, putting the fastening medium there into tension.

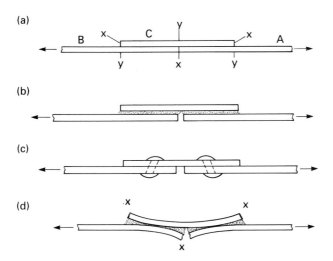

8.2 (a) Layout of single cover butt joint; (b) adhesive in shear; (c) fastener in shear; (d) bending of plates causing peeling at the joint's ends.

The mechanisms described above will occur, to a greater or lesser extent, in all joints. The ends of the joint will be the critical areas since it is here that the joining medium is most highly loaded, in a combination of shear and through-thickness tension (peel). Not only must the adhesive or fasteners be able to withstand these stresses, but also the plates being joined.

Because composite plates are relatively weak through the thickness, failure is often caused by the peel stresses; one of the main objectives in design, therefore, is to minimise peeling. This is readily achieved by tapering the plate thickness. More complicated joints, which will also induce shear and peel, can be improved by overall configuration; a few examples are shown in Fig. 8.3, which illustrates examples of good and bad practice in joint configurations.

8.4 Bonded joints

8.4.1 Introduction

Bonding could be regarded as the 'natural' way to fasten composite panels. Certainly it is the obvious approach for thin laminates. As already noted there are a number of disadvantages associated with bonding and these must be taken into account when selecting the approach to be used.

A satisfactory joint will only be achieved by adopting a consistent,

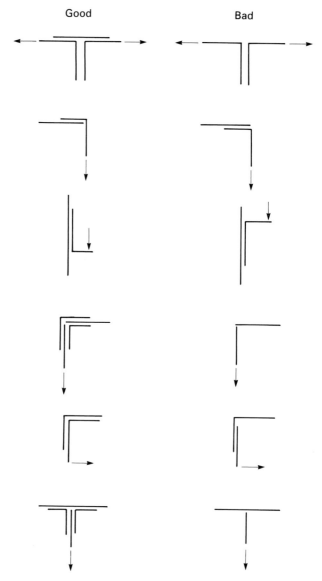

8.3 Examples of good and bad joint configurations.

and rigorous, approach throughout the manufacturing process. Particular attention must be paid to the surface preparation, the level of moisture in the workshop environment, the laminate and the adhesive, and to the selection of the appropriate adhesive. Finally the cure schedule recommended by the adhesive supplier must be followed.

Of the various parameters that determine the strength of a bonded joint, overlap length is one of the most important. It is false economy to make the overlap too short. The additional structural weight incurred by a longer overlap is more than compensated by the high achievable load levels in the structure as a whole, and by the resistance of the joint to creep and to flaws within the adhesive.

8.4.2 Surface preparation

Although some adhesives are less sensitive than others to surface contamination, unless the surfaces to be bonded are in a suitable condition, bonded joints will have a very low strength and/or durability. One approach is to incorporate into the laminate during manufacture a surface peel-ply. This ply is removed immediately before bonding and generally no further preparation is required. However, it is essential to ensure that no contaminants have been transferred from the peel-ply to the laminate's surface.

Laminates that do not incorporate a peel-ply will have their surfaces contaminated by the mould release agent. It is essential that all traces of contamination are removed as described below. Hand abrasion is not recommended.

1 Ensure composite is dry.
2 Blast area to be bonded (three passes) with dry alumina grit (280 grade).
3 Remove dust with clean running water.
4 Wipe bonding area with clean tissue.
5 Protect surface from contamination; keep composite dry.
6 Ensure adhesive has attained room temperature before bonding.
7 Bond as soon as possible after preparation, ideally in a facility with controlled humidity (<40%).

When bonding composite-to-metal adopt the following procedure:

1 Treat composite as above.
2 Treat metal as for metal-to-metal bonding (e.g. aluminium alloy should have standardised phosphoric acid or chromic acid etch).
3 Bond immediately.

8.4.3 Adhesive selection

Epoxy-based adhesives are the most commonly used for bonding composite-to-composite (especially CFRP) and composite-to-metal. For good environmental (creep and moisture) resistance, choose an adhesive which cures at elevated temperature. For good peel and fatigue resistance, choose a ductile adhesive. Final choice must be determined by joint configuration, required load capacity and service conditions.[8.1]-[8.3]

Characteristics of the basic systems are summarised in Table 8.2.

8.4.4 Adhesive suppliers

1   Bostik Ltd, Ulverscroft Road, Leicester LE4 6BW, UK.
2   Ciba-Geigy, Plastics Division, Duxford, Cambridge CF2 4QA, UK.
3   Cyanamid Fothergill Ltd, Abenbury Way, Wrexham Industrial Estate, Wrexham, Clwyd LL13 0NT, UK.
4   Dexter Corporation, Adhesives Division, 2850 Willow Pass Road, PO Box 312, Pittsburg, California 94565 - 0031, USA.
5   Dow Corning Ltd, Reading Bridge House, Reading RG1 8PW, UK.
6   Loctite (UK), Watchmead, Welwyn Garden City AL7 1JB, UK.
7   Lord Corporation (UK) Ltd, Stretford Motorway Estate, Barton Dock Road, Stretford, Manchester M32 0ZH, UK.
8   Permabond Adhesives Ltd, Woodside Road, Eastleigh SO5 4EX, UK.
9   3M United Kingdom plc, 3M House PO Box 1, Bracknell RG12 1JU, UK.
10   Narmco Materials Inc. 1440N Kraemer Blvd, Anaheim, California 92806, USA.

8.4.5 Joint geometry

8.4.5.1 Configurations

The strength of bonded joints is closely related to adherend thickness as illustrated in Fig. 8.4.[8.4] The weakest joints are those where failure is limited by interlaminar failure of the adherend or peel of the adhesive. The next strongest joints are those in which the load is limited by the shear strength of the adhesive. The strongest joints will fail outside the joint area at a load equivalent to the strength of the adherend.

Adhesive layers are at their most efficient in the thickness range 0.1–0.25 mm; thicker bonds are not practicable because of the impossibility of making them without unacceptable levels of flaws or porosity. An unsupported single lap joint is the weakest configuration and, for practical lay-ups, will never be as strong as the adherends being joined. Acceptable efficiencies can be achieved provided the overlap-to-thickness ratio is sufficiently large. Thicker adherends would need to be tapered at the ends of the overlap.

For thicknesses of greater than 1.5–1.75 mm for quasi-isotropic CFRP lay-ups (less for unidirectional laminates) a double-lap configuration is needed to transfer the strength of the adherends. The optimum overlap-to-thickness ratio is about 30 : 1. If the adherends are uniform,

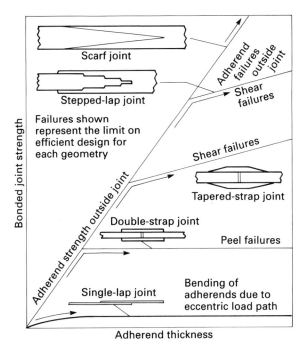

Scarf joint

Adherend failures outside joint

Shear failures

Stepped-lap joint

Failures shown represent the limit on efficient design for each geometry

Shear failures

Tapered-strap joint

Adherend strength outside joint

Double-strap joint

Peel failures

Single-lap joint

Bending of adherends due to eccentric load path

Bonded joint strength

Adherend thickness

8.4 Effect of adherend thickness on bonded joint strength (after ref. [8.4]).

thickness is limited to about 4.5 mm, whilst tapering can increase the limit to 6.35 mm.

For thicknesses greater than 6.35 mm stepped or scarf joints should be considered. Theoretically, merely by making the angle small enough, it should be possible to make a scarf joint stronger than the adherend being joined. In practice, particularly for wide joints, it will be impossible to make the required knife-edge. Also, for very thick adherends the length of the joint may be prohibitive.

8.4.5.2 Dimensions
Joint dimensions are defined in Fig. 8.5. Real joints will not display the square-ended adhesive layer illustrated because of the formation of a 'spew-fillet'. The presences of this fillet enhances joint strength.

8.4.6 Failure modes

The possible failure modes are illustrated in Fig. 8.6. An alternative mode of failure at the adherend surface/adhesive interface is indicative of inadequate preparation and should be treated as a quality control problem.

Table 8.2 Characteristics of adhesive systems

Type	Form	Cure temp. (°C)	Maximum service temp (°C)	Advantages	Disadvantages	Example (Supplier)*
Epoxy	Paste or liquid	Room or elevated	80–100	Easy storage; easy mixing and use; long shelf life; gap filling	Poor/moderate environmental resistance; low peel strength	AY 103 (2) Permabond EO4, E34 (8) Araldite 2001 (2) EC 1838 B/A (9)
	Paste or liquid	100–200	120–160		Improved environmental resistance; low peel strength unless toughened	Permabond ESP110 (8) EC3448, AF163 (9) EA9391 (4)
	Film	120–170	120–180	Wide range of properties; controlled thickness; good environmental resistance	Needs refrigerated storage; can be brittle	FM300 (3) Rodux 312, 319 (2) Metlbond 329 (10) EA9628 (4) AF191 (9) BSL308 A (2)
Acrylic	Paste or liquid	Room or elevated	120–180	Fast setting; easy to use; good environmental resistance; tolerant of surface contamination; used for GRP	Limited pot life	Permabond F241 (8) Multibond (6) Versilok (7) Hyperbond (1)
Polyurethane	Liquid	Room or elevated	80	High peel strength; gap filling; cryogenic use; used for GRP	Moisture-sensitive	Permabond 8000 (8) 7501 A/B (7) XB5000 series (2)
Silicone	Paste	Room	260	High peel strength; good impact resistance	Poor shear strength	Silastic 738 (5)

| Polyimide | Film | 175–280 | 300 | Good high temperature performance | Low peel strength: high temperature cure and post-cure | FM35, FM456 (3) |
| Phenolic | Film | 180 | 250 | High temperature use; good solvent resistance | Brittle | HT424 (5) |

* Numbers in brackets given in last column refer to suppliers listed in Section 8.4.4.

Adherend tensile failure will occur as the result of in-plane tensile and, possibly, bending stresses. Transverse adherend failure is the result of Poisson's ratio contractions and is often seen in adherends rich in 0° plies (parallel to the load). Interlaminar adherend failure will be caused by in-plane through-thickness shear and/or through-thickness tension (i.e. peel) stresses. Adhesive failure will be in shear or peel. In general, a ductile adhesive is to be preferred to a brittle (and stronger) alternative.

8.4.7 Adhesive characteristics

Any calculations to determine the strength of a bonded joint will require data on the mechanical properties of the adhesive to be used. In particular the shear stress–shear strain curve should be known. Such information, if not available from the adhesive supplier, may sometimes be obtained from ESDU (Engineering Sciences Data Unit, 27 Corsham Street, London N1 6UA). Alternatively the properties can be obtained from a 'napkin-ring' or 'thick adherend' test.

Brittle adhesives, which show virtually linear (elastic) shear stress–strain curves to failure, have a shear failure strain of around 0.1. Ductile adhesives have an initial linear characteristic followed by a plastic

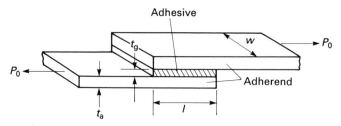

8.5 Definitions of joint dimensions: t_a, adherend thickness; t_g, adhesive thickness; l, overlap length; w, joint width; P_0, applied load.

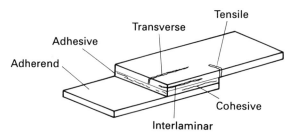

8.6 Possible failure modes in bonded joints between composite adherends.

region of high strain with a much reduced modulus; failure strain can be as high as 2.0.

For the purposes of calculation it is convenient to represent the actual shear stress–strain curve by an equivalent bilinear representation, as shown in Fig. 8.7.[8.5] A typical brittle adhesive would have a ratio $\gamma_p/\gamma_e = 1.5$, and a ductile adhesive a ratio of 20. The actual curve used should correspond to service temperature and moisture conditions.

8.4.8 General design considerations

1 Both shear and peel stresses have their maxima at the ends of the overlap in single and double lap joints or the ends of a step in a stepped joint. Stresses will be a minimum in the centre of these regions. The shear stress is essentially constant along a scarf joint.
   Maximum stresses are reduced by:
   (a) using identical adherends (if not possible, equalise in-plane and bending stiffness, and coefficients of thermal expansion);
   (b) using highest possible in-plane laminate stiffness;
   (c) using longest possible overlap;
   (d) using adhesive with low tensile and shear elastic moduli.
2 Taper the adherends to reduce maximum peel stresses.
3 Calculate the bond shear strength to have a margin of 50% above the adherend strength; this reserve protects against environmental degradation and fatigue loading.
4 Choose a ductile adhesive; this gives a higher static and fatigue strength.

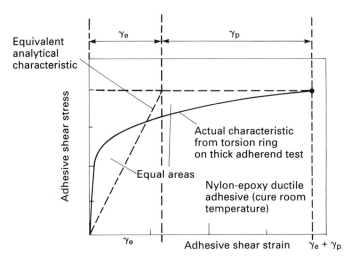

8.7 Elastic–plastic representation of actual adhesive shear characteristics (after ref. [8.5]).

5 Ensure that the minimum adhesive shear stress does not exceed 10% of τ_p; this will allow accumulated creep strains to recover.
    Creep can only be avoided in a scarf joint by limiting the maximum strain.
6 Use a 'homogeneous' rather than a 'blocked' stacking sequence (i.e. do not group together plies of the same orientation).
7 For fatigue (cyclic) loading, a life of 10^6 cycles can be expected if the peak stress is limited to about 25% of the static strength value for single lap joints, and to about 50% for scarf joints.

8.4.9 Design of a single lap joint

The design of single lap joints is discussed in refs [8.5] and [8.6].

8.4.9.1 Strength
For long overlaps ($50 < l/t < 100$) the strength of single lap joints is only weakly dependent on the strength of the adhesive (especially if this is ductile).

The maximum longitudinal stress in the adherend (adjacent to the bond line at the end of the overlap) for an applied load P_0 is given by

$$\sigma_{max} = \sigma_{av} + \sigma_{bend}$$

i.e. $$\sigma_{max} = \frac{P}{t} + \frac{6M_0}{t^2}$$

where $P = \dfrac{P_0}{w}$;

$$M_0 \approx \frac{Pt}{2[1 + \xi^c + 1/6(\xi^c)^2]} ;$$

$$\xi^c = \frac{P}{D} \text{ (D being the adherend bending stiffness);}$$

$$c = \frac{l}{2} .$$

The bending stress (σ_{bend}) is reduced (i.e. $\sigma_{max} \rightarrow \sigma_{av}$) as the bending moment M_0 is reduced (i.e. as l is increased or D is decreased). For isotropic adherends this is illustrated in Fig. 8.8.

8.4.9.2 Joint efficiency
If a single lap joint is restrained against rotation (by a supporting substructure) its efficiency will equal one half that of a double lap joint (see Section 8.4.10). For compressive shear loading (as opposed to tensile shear) support against rotation is essential.

For unsupported single lap joints acceptable efficiency will only be

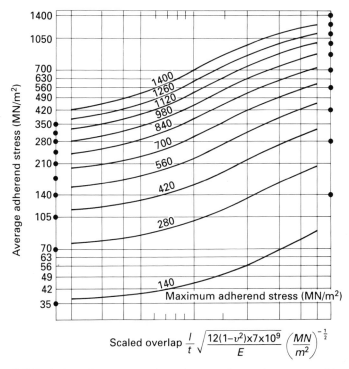

8.8 Variation of average and maximum adherend stresses with overlap length (after ref. [8.6]).

obtained with large overlaps ($l/t > 50$). Representative curves are given in Fig. 8.9–8.11. These curves relate to uniform thickness carbon–epoxy laminates, operating at room temperature, with the composite and adhesive properties given below.

Composite: E_{11} = 145 GN/m², E_{22} = 12 GN/m², σ_{11} = 1240 MN/m², σ_{22} = 55 MN/m².
Ductile adhesive: τ_p = 41 MN/m², γ_p = 1.95, γ_e = 0.098.
Brittle adhesive: τ_p = 62 MN/m², γ_p = 0.063, γ_e = 0.042.

(See Section 8.4.7 for definitions):

8.4.9.3 Miscellaneous issues

1 *Stacking sequence*: the bending stiffness of composite adherends can be altered independently of the in-plane stiffness (by changing the stacking sequence whilst keeping the lay-up constant). In general joint performance is improved if 0° plies (i.e. fibres in the load direction) are placed on the surface of the laminate; the overall effect is illustrated by the factor k_b in Fig. 8.12. However, there is

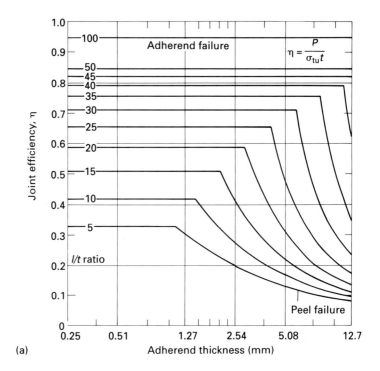

(a)

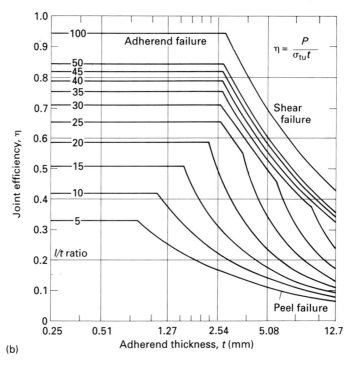

(b)

8.9 Joint efficiency charts (after ref. [8.6]): (0°/45°/–45°/90°) high tensile strength (HTS) carton-epoxy. (a) Ductile adhesive; (b) brittle adhesive (room temperature).

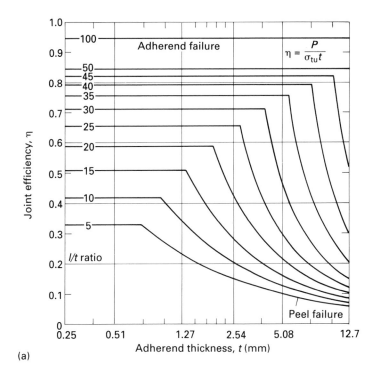

(a)

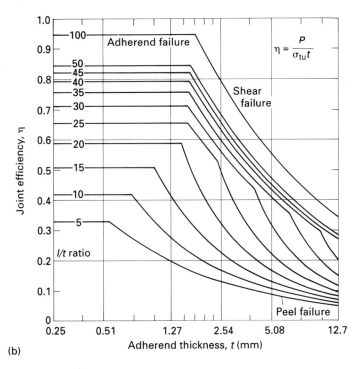

(b)

8.10 Joint efficiency charts (after ref. [8.6]): (0°/45°/−45°/0°) HTS graphite–epoxy. (a) Ductile adhesive; (b) brittle adhesive (room temperature).

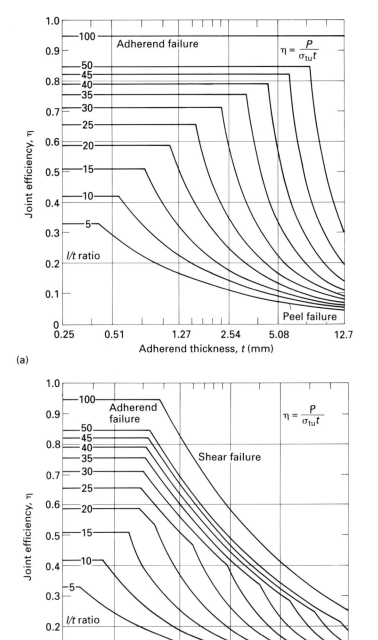

8.11 Joint efficiency charts (after ref. [8.6]): (0°) HTS graphite-epoxy. (a) Ductile adhesive; (b) brittle adhesive (room temperature).

evidence that use of k_b alone is not sufficient to place laminates in ranking order.[8.7]

2 *Non-identical adherends*: joint strength will always be reduced if the adherends are not identical. Reductions will be caused by stiffness imbalance (Fig. 8.12) or thermal coefficient mismatch. In the latter case general results cannot be obtained. In the event of no stiffness imbalance, thermal mismatch will increase the bending moment (at the end of the overlap) in the adherend with the smaller expansion coefficient.

3 *Tapering*: the tendency to fail by peel (for lower values of l/t – higher t or lower l) is reduced by tapering the adherends. The change of slope of the curves in Fig. 8.9–8.11 indicates the thickness above which adherends should be tapered.

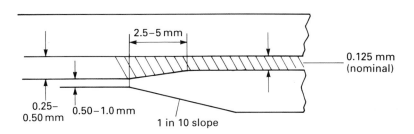

Further improvement can also be gained by thickening the adherend in the lap region.

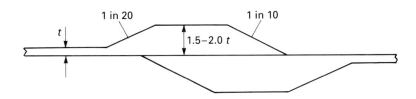

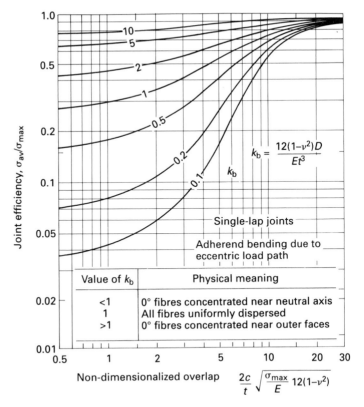

8.12 Effect of stacking sequence on joint efficiency: stiffness balanced joints (after ref. [8.6]). Note: c is defined on page 212.

8.4.10 Design of double lap joints

The design of double lap joints is discussed in refs [8.5] and [8.8].

8.4.10.1 Adhesive shear stress distribution
The load capability of double lap joints is strongly related to the distribution of shear stress in the adhesive. As shown diagrammatically in Fig. 8.13 most of the load is transferred at the ends of the overlap. The regions of plastic flow reach a constant value so that above a certain overlap no further load is carried.

To allow for recovery of creep strains, the minimum stress at the centre of the joint should not exceed $\tau_p/10$.

8.4.10.2 Strength
As explained in Section 8.4.10.1 the strength of double lap joints does not increase if overlap length is increased above a minimum length, given by:

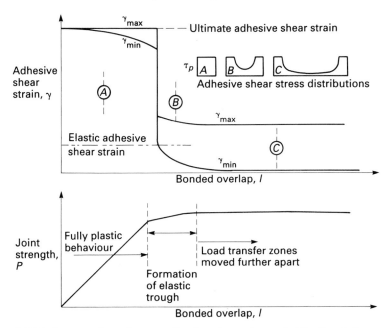

8.13 Influence of overlap on adhesive shear stress distribution (after ref. [8.4]).

$$l_{min} = \frac{P}{2\tau_p} + \sqrt{\left[\frac{\tau_p}{t_g \gamma e} \left(\frac{1}{E_o t_o} + \frac{2}{E_i t_i} \right) \right]}$$

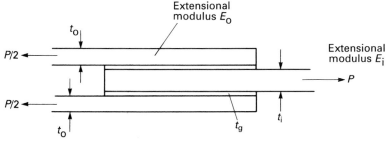

where P is the required joint strength per unit width. Normally this would be the strength of the adherend. For a stiffness balanced joint, $2E_o t_o = E_i t_i$.

The joint strength for tensile loading (accounting for both stiffness and thermal mismatch) may be taken as the smaller of:

$$P_1 = (\alpha_o - \alpha_i) \Delta T E_i t_i + \sqrt{\{2k\tau_p t_g (\gamma_e + \gamma_p) 2E_i t_i [1 + E_i t_i / 2E_o t_o]\}}$$

and

$$P_2 = (\alpha_i - \alpha_o) \Delta T 2 E_o t_o + \sqrt{\{2k\tau_p t_g (\gamma_e + \gamma_p) 4 E_o t_o [1 + 2 E_o t_o / E_i t_i]\}}$$

where α is the coefficient of thermal expansion; the temperature difference $\Delta T = T_{operating} - T_{cure}$, T_{cure} being the cure temperature of the adhesive; and $k = (\tfrac{1}{2}\gamma_e + \gamma_p)/(\gamma_e + \gamma_p)$.

If P_1 or P_2 is calculated to be negative the joint will fail, without application of an external load, because of excessive residual stresses arising from thermal mismatch.

If the loading is compressive the above procedure is repeated but with the sign of ΔT reversed. If the loading is (edgewise) shear, extensional stiffness Et is replaced by shear stiffness Gt of the adherends and the thermal term ignored.

Joint strength of thicker adherends is limited by peel stresses. The maximum peel stress, which occurs at the end of the overlap, is given by:

$$\frac{\sigma_{peel}}{\tau_p} = \left(\frac{3 E_g (1 - v^2) t_o}{E_o t_g} \right)^{1/4}$$

where v is the adherend Poisson's ratio and E_g is the tensile modulus of the adhesive. σ_{peel} should be equated with the smaller of the adhesive tensile strength or (more usually) the interlaminar tensile strength of the adherend.

8.4.10.3 Efficiency

The efficiency of double lap joints is illustrated for CFRP in Fig. 8.14. The changes in slope indicate the thickness above which peel stress reduction by tapering (see Section 8.4.9.3) is indicated, or above which a scarf or stepped joint should be employed due to shear strength limitations. The adherend and adhesive properties used for Fig. 8.14 are the same as those used for Fig. 8.9–8.11.

8.4.11 Design of stepped joints

The design of stepped lap joints is discussed in refs [8.5] and [8.9].

8.4.11.1 Stress distribution and analysis

The adhesive shear stress distribution, within a step, is similar to that in a double lap joint (see Section 8.4.10.1). It follows that there will be a limit to the load that can be transferred at any step.

For prediction of joint strength it is essential to use a stress analysis method that allows for the non-linear stress–strain properties of the adhesive.[8.4] Computer software can be obtained from such organisations as ESDU (Engineering Sciences Data Unit, 27 Corsham Street, London N1 6UA).

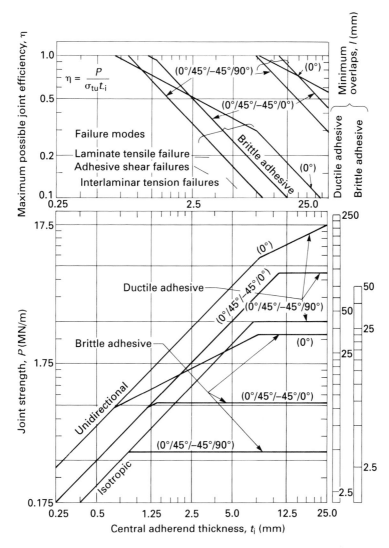

8.14 Efficiency and strength chart for CFRP double lap joints (after ref. [8.8]): HTS graphite-epoxy (room temeprature).

8.4.11.2 General guidelines

1 Stepped joints are normally used to bond composite to metal. Ideally they should be symmetric as shown in Fig. 8.15.

2 Stiffness should be constant along the joint, i.e.

$$E_{metal} t_{metal} + E_{comp} t_{comp} = \text{constant}$$

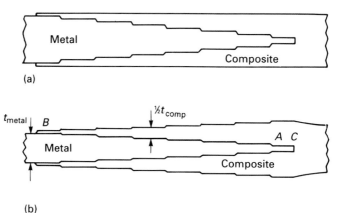

8.15 Stepped lap joints: (a) constant thickness (stiffness unbalanced); (b) constant thickness (variable thickness).

3   Length of end step in metal (*A*, Fig. 8.15) should not exceed 10 mm; thickness should not exceed 0.75 mm.
4   Length of end step in composite (*B*, Fig. 8.15) should not exceed 13 mm; (half) thickness should not exceed 0.75 mm.
5   Intermediate steps should be about 13 mm long, with thickness increment of 0.5–0.75 mm.
6   Central step should be at least 25 mm long (to allow for relief of creep strains).
7   Thickness increment should match ply thicknesses (maximum joint strength with a separate step for each ply).
8   Fibres adjacent to the metal should be oriented along the joint.
9   Plies should not stop short of a step, nor overlap on to the next step.
10   Central plies should be diverted around a low modulus triangular wedge, about 13 mm long (*C*, Fig. 8.15).
11   If it is necessary to make a single-sided (non-symmetric) joint, a cold-setting adhesive, or thermally compatible adherends, should be used to avoid warping.

8.4.12 Design of scarf joints

The design of scarf joints is discussed in refs [8.5] and [8.9].

8.4.12.1 Ideal joint
The design of scarf joints should be considered with reference to an 'ideal' configuration, i.e. a joint between identical adherends with a perfectly sharp tip to each adherend.

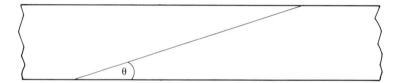

In these circumstances, for a long overlap ($\theta \leqslant 5°$), the adhesive shear stress is effectively uniform. The strength of the joint can be calculated from the equation relating adherend axial stress (σ) to adhesive shear stress (τ), i.e.

$$\sigma \sin\theta = \tau$$

8.4.12.2 Practical considerations

1   If a perfectly sharp tip cannot be maintained in practice, a finite thickness of about 0.25 mm should be used. In this case the joint should be analysed as for a stepped joint.
2   If adherends are not identical, the effect of stiffness mismatch can be estimated from Fig. 8.16 and the effect of thermal mismatch from Fig. 8.17. These figures are derived from elastic analyses; the use of ductile adhesive will give improved performance.
3   Thermal mismatch is minimised by using a multiple scarf configuration.

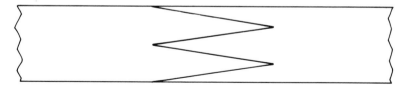

4   To minimise creep deformation the shear stress in the adhesive must be limited (for the ideal case the stress is uniform along the joint).

8.5 Mechanically fastened joints

8.5.1 Introduction

Compared with bonded joints, mechanically fastened joints are relatively easy to prepare. However, care must be taken when drilling holes, and accurate fit of all fasteners is essential if the design load is to be achieved. Correct fastener selection is vital if high efficiency is to be obtained; special fasteners are now produced by a number of suppliers.

Because a mechanically fastened joint can never be as strong as the laminate being joined, increased thickness in the joint region will be

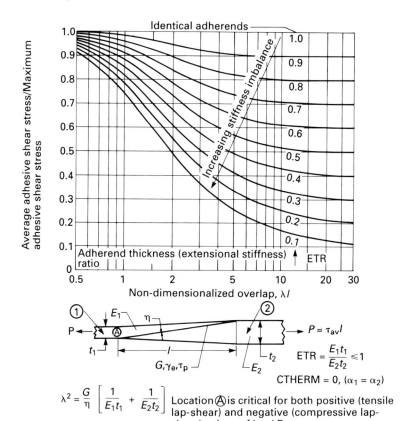

8.16 Effect of adherend stiffness imbalance on strength (after ref. [8.9]). (ETR – ratio of product of adherend thickness and extensional stiffness E_1t_1/E_2t_2, (THERM – thermal mismatch of adherends.)

required. The weight penalty will be compensated by the increase in overall structural efficiency.

8.5.2 Hole preparation

8.5.2.1 Hole tolerance
Unless otherwise specified by fastener suppliers, holes should be a 'net' fit (0 ± 0.1 mm).

8.5.2.2 Drills
High speed steel bits are not recommended. Carbide or diamond coated tools should be used.

Points should be sharper than for metals: 55–60° for thin sheets, 90–100° for thicker sheets.

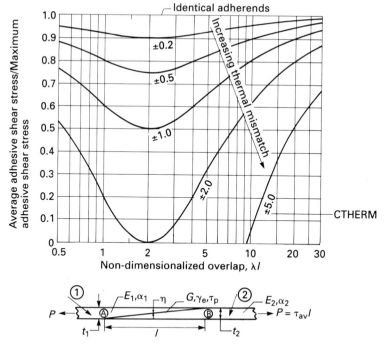

$$\lambda^2 = \frac{G}{\eta}\left[\frac{1}{E_1 t_1} + \frac{1}{E_2 t_2}\right] \quad \text{CTHERM} = \frac{(\alpha_2-\alpha_1)\Delta T \lambda}{\tau_p\left(\frac{1}{E_1 t_1} + \frac{1}{E_2 t_2}\right)} \quad \Delta T = T_{operating} - T_{stress\,free}$$

Location A critical for CTHERM < 0 and P > 0
Location A critical for CTHERM > 0 and P < 0
Location B critical for CTHERM < 0 and P < 0
Location B critical for CTHERM > 0 and P > 0

8.17 Effect of adherend thermal mismatch on strength (after ref. [8.9]).

8.5.2.3 Drilling

Where possible a sacrificial backing sheet should be used. Otherwise use of a controlled-feed drill is recommended. Lubricant fluids should not be used, but dust extraction is essential.

Frequent backing-off, to remove debris, is recommended. Surface cutting speeds of around 20 mm/min and feed rate of 10–20 mm/min are recommended.

When countersinking, a combined drill/countersink tool, with initial pilot hole, should be used. The countersink depth should not exceed one half the laminate thickness.

8.5.2.4 Materials
GFRP and CFRP can be treated similarly. Care must be taken with CFRP to avoid exit surface splintering.

Kevlar fibre reinforced polymers (KFRP) will need special drill bits. A surface layer of 120 style glass-cloth will improve hole quality.

8.5.2.5 General
Maintenance of hole tolerance, circularity and position are essential to ensure even fastener loading in multifastener joints.

8.5.3 Fastener selection

8.5.3.1 Screws
Self-tapping screws are not recommended for load-carrying joints; they may be used as anti-peel fasteners in bonded joints.

8.5.3.2 Rivets
Conventional rivets (as used for joining metal) are not recommended. Special-purpose rivets giving controlled expansion and clamp-up, with suitable head geometry, should be used.

Laminates giving a combined thickness of 6–8 mm can be satisfactorily joined.

8.5.3.3 Bolts
Conventional (uncountersunk) bolts can be used; they should be tightened to the standard torque and the nut should be locked to prevent loss of clamping in service. The use of special-purpose bolts is recommended for flush-fitting and blind installation.

8.5.3.4 General
To avoid corrosion, the correct material must be chosen when fastening CFRP.

8.5.4 Fastener suppliers

1 Ateliers Haute Garonne, Zone Industrielle, 31130 Flourens, Toulouse, France. Flush metallic rivets ('FYBRFAST').
2 Camloc (UK) Ltd, 15 New Star Road, Leicester LE4 7JD, UK. Metallic panel fasteners and inserts ('TRIDAIR', 'CLICK BOND').
3 CJ Fox & Sons Ltd, Shoreham Airport, Shoreham by Sea BN43 5FN, UK. Protruding and flush head metallic and composite bolts ('CHERRY ACP', 'E-Z-BUCK', etc.).
4 Huck-UK Ltd, Unit C, Stafford Park 7, Telford TF3 3BQ, UK. Various metallic bolts and rivets ('HUCK-COMP', HUCK-TITE', etc.).
5 Monogram Aerospace Fasteners, 3423 South Garfield Avenue, Los

Angeles, California, 90022 - 0547, USA. Protruding and flush head blind metallic bolts ('COMPOSI-LOK').
6   SPS Technologies, Highland Avenue, Jenkintown, PA 19046, USA. Protruding head, flush and blind metallic bolts ('COMP-FAST' and 'COMP-TITE').
7   Tappex, Masons Road, Stratford-upon-Avon CV37 9NT, UK. Threaded inserts.
8   Tiodize, 5858 Engineer Drive, Huntingdon Beach, California 92649, USA. Protruding and flush composite bolts ('FIBERLITE').

8.5.5 Joint geometry

8.5.5.1 Configurations
The same configurations as for bonded joints can be used – single lap, double lap, stepped, scarf (see Section 8.2) Where possible the loading should be symmetrical, i.e. double shear.

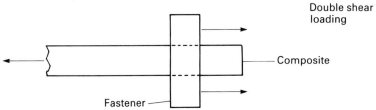

Where laminates are tapered, it is essential to use matching tapered washers. Some special fasteners can accommodate limited blind-side slope (e.g. 'COMPOSI-LOK' bolts).

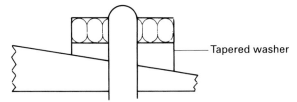

8.5.5.2 Dimensions
Joint dimensions (defined in Fig. 8.18) are determined according to the material being fastened (see Section 8.5.8). The ratio of d/t should lie within the range 1.5–3.0.

8.5.6 Failure modes

The possible failure modes are illustrated in Fig. 8.19. Shearout is avoided by having adequate end distance ($> 4d$). Cleavage is avoided by having adequate width and end distance ($w, e > 4d$). Pull through is likely only with countersunk heads. Use the maximum possible head angle (120° or 130°). Fastener shear failure is avoided by selecting correct

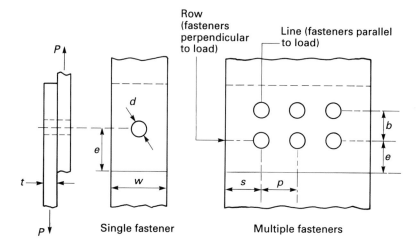

8.18 Definition of joint dimensions: t, plate thickness; d fastener diameter; w, joint width; e, end distance; s, side distance; p, fastener pitch; b, back pitch; P, applied load.

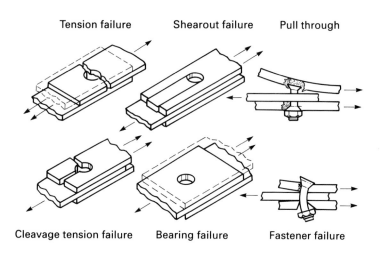

8.19 Joint failure modes.

diameter; bending failure is usually avoided if $d/t > 1$. Tension and bearing failure are essentially material-dependent (see Section 8.5.8). Bearing failure is usually benign (non-catastrophic); tension failure is generally catastrophic.

8.5.7 Laminate material

Laminate materials are discussed in refs [8.10] and [8.11].

8.5.7.1 Fibre type
General characteristics are not affected by fibre type. Strength levels are approximately in the ratio 1 : 0.8 : 0.5 for laminates with carbon (C), glass (G), or Kevlar (K) fibres respectively.

8.5.7.2 Laminate lay-up
Optimum performance is obtained with a quasi-isotropic $(0/90/\pm45°)$ lay-up; in any one direction fibre proportions should lie in the range 12.5–37.5%. For $(0/\pm45°)$ lay-ups, the $\pm45°$ fibre proportions should lie in the range 37.5–75%.

8.5.7.3 Stacking sequence
Stacking sequence is of minor importance for fully tightened bolts. For rivets, stacking sequence is more important; a 90° or 45° surface layer is recommended (load in 0° direction).

Plies of the same orientation should not be grouped together; a change in fibre orientation from one ply to the next is recommended.

8.5.8 Static ultimate strength of single hole bolted joints

8.5.8.1 Strength charts
The charts (Fig. 8.20–8.23) give, for a range of composites, the variation of static strength with w/d and e/d ratios. It is seen that the value of ultimate strength, which is given in terms of the bearing stress, σ_b (= P_{ult}/dt, P_{ult} being the load at failure), reaches a 'plateau'. For low values of w/d the joint fails in tension and for low values of e/d the joint fails in shear.[8.10, 8.12–8.15]

The 'knee' in the curves represents the change in failure mode from tension or shear to bearing. Although the plateau represents pure bearing failure, this mode generally occurs before the plateau is reached.

The quoted values are for single hole joints in tension in double shear via fully tightened protruding head bolts, with a plain shank through the full thickness of the laminate, which is clamped between the washers beneath head and nut. The effect of changing d/t ratio and of bolt tightening are described later (Sections 8.5.8.3 and 8.5.9). Fibre volume fraction, v_f, can be taken as 0.60 unless otherwise indicated. The laminate is dry. The load is always parallel to the 0° direction. The

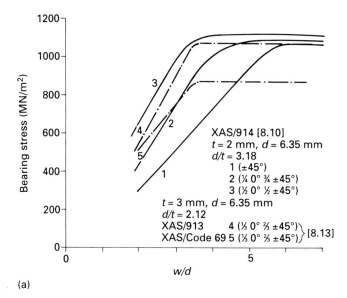

(a)

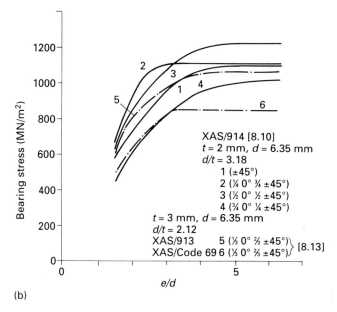

(b)

8.20 Variation of bearing strength with (a) *w/d* and (b) *e/d* for CFRP.

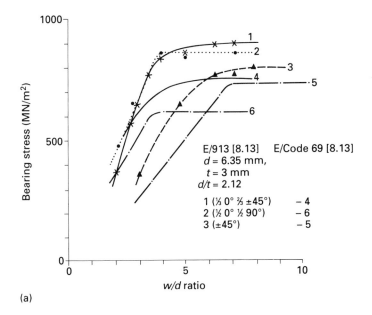

(a)

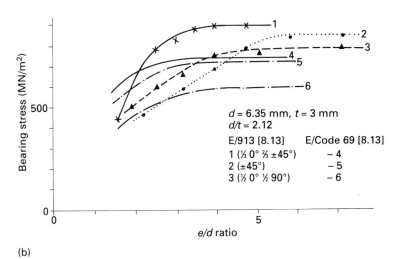

(b)

8.21 Variation of bearing strength with (a) w/d and (b) e/d for GFRP.

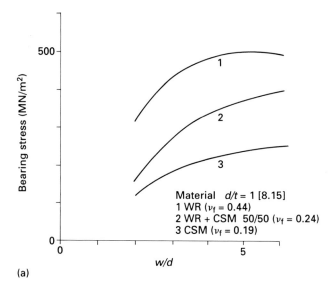

(a)

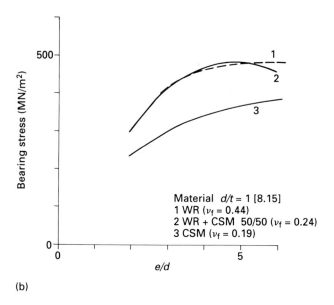

(b)

8.22 Variation of bearing strength with (a) w/d and (b) e/d for GRP.

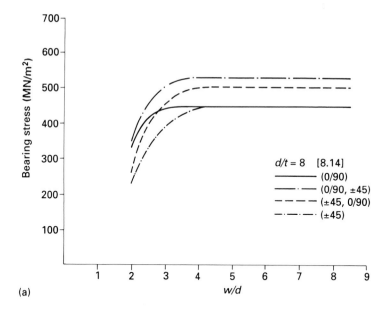

(a)

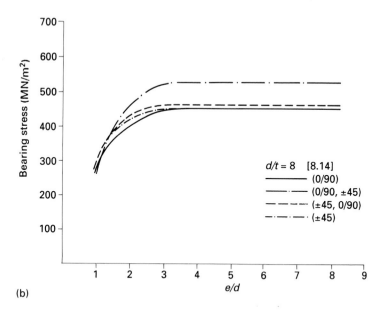

(b)

8.23 Variation of bearing strength with (a) w/d and (b) e/d for woven bidirectional KFRP.

curves represent mean values; typically one standard deviation would be about 10% of the mean.

8.5.8.2 Calculation of ultimate bearing strength

The single fastener strength (full-tightened in double shear) can be calculated for CFRP using the method of Collings[8.4] as given below. The same approach is expected to work for GFRP. There is no evidence of its applicability to KFRP.

Lay-up	Bearing strength
$0/\pm45°$	$\dfrac{1}{t}\left(\dfrac{100\,t_o\,\sigma_c\,\sigma_{bo}}{\sigma_c\,(100-\phi_{45})+\phi_{45}\,\sigma_{bo}}+t_{45}\,\sigma_{b45}\right)$
$90/\pm45°$	$\dfrac{1}{t}\left(\dfrac{100^2\,t_{90}\,\sigma_{TC}}{100^2+2.38\,\phi_{90}^2}+t_{45}\,\sigma_{b45}\right)$
$0/90°$	$\dfrac{1}{t}\left(\dfrac{100^2\,t_{90}\,\sigma_{TC}}{100^2+2.38\,\phi_{90}^2}+t_o\,\sigma_{bo}\right)$
$0/90/\pm45°$	$\dfrac{1}{t}\left(\dfrac{100\,t_o\,\sigma_c\,\sigma_{bo}}{\sigma_c(100-\phi_{45})+\phi_{45}\sigma_{bo}}+\dfrac{100^2\,t_{90}\,\sigma_{TC}}{100^2+2.38\phi_{90}^2}+t_{45}\,\sigma_{b45}\right)$

where t	= laminate thickness;
t_0, t_{90}, t_{45}	= thickness of $0°, 90°, \pm45°$ plies respectively;
ϕ_{45}, ϕ_{90}	= percentage of $\pm45°, 90°$ plies respectively;
σ_c	= $0°$ longitudinal compression strength;
σ_{TC}	= constrained transverse compression strength;
σ_{bo}	= constrained $0°$ bearing strength;
σ_{b45}	= constrained $\pm45°$ bearing strength.

8.5.8.3 Effect of varying d/t

Bearing strength varies with d/t as indicated in Fig. 8.24. The strength changes are more marked for bolts that are not fully tightened; such data would apply also to pins and rivets. However, the ultimate bearing strength of finger-tight bolts or pins is significantly less than that of fully tightened bolts being typically 50% of the fully tightened value for CFRP[8.10] and 75% for GFRP.[8.13]

8.5.9 Riveted joints

Very few data are available for riveted joints. The variations of strength with w/d and e/d are similar to those seen for bolts. Most published results are for single lap joints and it is clear that strengths are very sensitive to rivet type (hollow or solid) and head form (protruding or countersunk). The depth of countersink should be limited to one half the thickness if a drastic loss of strength is to be avoided.

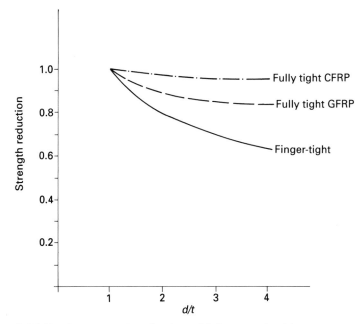

8.24 Bearing strength reduction with increase in d/t.

For thicker laminates failure by rivet shear is likely. Failure loads should be checked using manufacturers' strength data.

Typical values of ultimate static bearing strength are given in Table 8.3 for single lap joints. Variation with d/t will be similar to that shown in Fig. 8.24. The quoted values can be increased by 50% for double lap joints.

8.5.10 Multifastener joints

Arrays of fasteners are composed of rows (perpendicular to the load) and lines (parallel to the load); see Fig. 8.18. Unless the pitch is sufficiently large, the strength per fastener in a row joint will be less than that of a single hole joint. In such cases the joint strength, P, is given by the expression:

$$P = nKP_1$$

where n is the number of fasteners, K is taken from Fig. 8.25 and P_1 is the single fastener strength for the appropriate value of w/d. The s/d (see Fig. 8.18 for definition of s) should be adequately large (take $2s = w$).

In line joints there will be an interaction between the bearing load (P_{br}) and the by-pass load (P_{bp}) (introduced by other fasteners).

Table 8.3 Ultimate static bearing strength of rivets

Material	Lay-up	Rivet	D (mm)	d/t	σbult (MN/m²)
Carbon XAS/914	±45 °	S/P	3.17	3.17	515
		S/C	3.17	3.17	420
		H/P	3.17	3.17	420
		H/C	3.17	3.17	340
	$0_2/\pm45$ °	S/P	3.17	3.17	470
		S/C	3.17	3.17	380
		H/P	3.17	3.17	405
		H/C	3.17	3.17	350
Glass E/DX/210	0/90 °	S/P	3.17	3.17	300
		H/P	3.17	3.17	250
Kevlar woven	±45 °	S/P	3.17	3.17	410
fabric/913		H/P	2.78	2.78	205
		H/P	3.17	3.17	228
	0/90 °	S/P	3.17	3.17	364
		S/P	3.17	3.17	580
		H/P	3.17	3.17	230
		H/C	3.17	2.12	250
	0/90/±45 °	S/P	3.17	3.17	366
		H/P	3.17	3.17	250

S, solid rivet; H, hollow rivet; P, protruding head; C, 120° countersunk head.

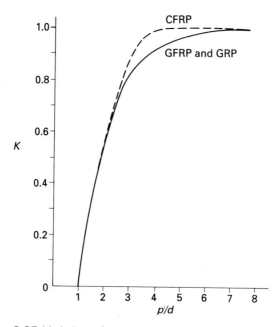

8.25 Variation of strength reduction factor K with pitch/diameter ratio.

Strength will be reduced, as for rows, if fasteners are too close. Multiple lines can be treated as single lines if the spacing is adequate ($K \rightarrow 1$).

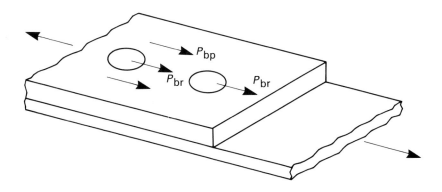

In a balanced two-fastener line (load shared equally), joint strength, P_2, will be typically only 10% greater than that, P_1, of the otherwise single fastener joint, i.e.

$$P_1 = \hat{\sigma}_b dt$$
$$P_2 = 1.1 \, (2 \, \sigma_b \, dt)$$

where $\sigma_b = \frac{1}{2} \hat{\sigma}_b$.

To obtain further increase in strength (up to 40% maximum) a line of (at least) four fasteners will be needed. A comprehensive analysis, including material properties and fastener flexibilities, is needed to determine the load on each fastener. Plate thicknesses and fastener diameters will have to vary along the length of the joint in order to minimise the bearing load on the most critical fastener. Such complexity is only justified for highly loaded joints, usually between composite and metal.[8.11] Generally, increasing the local laminate thickness and using one or two rows of fasteners is to be preferred.

8.5.11 Miscellaneous issues

8.5.11.1 Load direction
Static bearing strength is insensitive to load direction (relative to $0°$ plies) for fully tightened fasteners in other than highly orthotropic lay-ups.

For finger-tight bolts, pins and rivets, a reduction of about 15% in bearing strength should be expected in a (50% $0°$, 50% $\pm45°$) lay-up as the load direction changes from $0°$ to $90°$. The latter would correspond to a joint loaded in edgewise shear.

For loading in compression, bearing strength would be increased by 10% for CFRP and GFRP with moderate values of w/d and e/d.

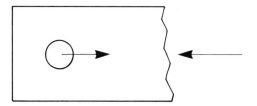

8.5.11.2 Fatigue loading

For net fit fasteners, joints show a reduction in strength of about 10% after 10^6 cycles of fatigue loading. It is vital that there are no loose fit fasteners in a multihole joint; strengths will be reduced below that just quoted if a fastener becomes loose in its hole.

8.4.11.3 Environmental effects

Bearing strength is reduced from the values previously quoted (Fig. 8.20–8.23) if the laminate is hot and/or wet. The basic double shear strength should be multiplied by the factor given in Table 8.4 (based on tests on CFRP).

Table 8.4 Factors for different operating conditions

Operating condition	Factor
Room temperature, wet	0.88
127 °C, dry	0.70
127 °C, wet	0.60

8.5.11.4 Single shear loading

For bolted single shear loading (i.e. single lap joints) multiply the double shear strength by the factor given in Table 8.5.

Table 8.5 Factors for single shear loading

	Single fastener or single row	Two in-line fasteners or two rows
CFRP	0.90	0.93
GFRP	0.85	0.90
KFRP	0.80	0.85

8.5.11.5 Countersunk fasteners

The bearing load should be calculated on the depth of the plain shank only: hence $P = \sigma_b d t_p$ with $t_p \geqslant \frac{1}{2}t$.

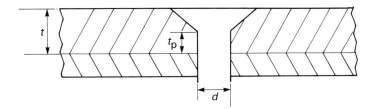

8.5.11.6 Lap configuration

The strengths quoted previously (Fig. 8.20–8.23) would apply to the outer plate of a double lap joint. For such a joint the inner lap is found to have a bearing strength 10% higher than these values.

To obtain a balanced joint it is recommended that the outer laps have a thickness $t_o = 0.55\, t_i$, for $1.5 < d/t < 3.0$. If $1.0 < d/t < 1.5$, then $t_o = 0.6\, t_i$.

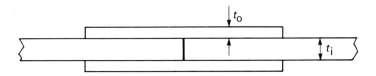

8.5.11.7 Hole elongation/allowable stress

Damage occurs to the laminate, in the vicinity of fastener holes, at a relatively low fraction of ultimate load. To avoid excessive damage, and associated hole extension, under static load, or hammering under fatigue loading, the following allowable stresses are recommended:

for CFRP:	75% of ultimate bearing
for GFRP or KFRP:	50% of ultimate bearing

8.5.12 Design procedure for bolted joints

8.5.12.1 General

Designers should try to avoid a laminate lay-up that is highly orthotropic. Where this is unavoidable, for example when seeking to achieve high stiffness or strength in a certain direction, then it may be necessary to modify the lay-up, by introducing additional plies, in the joint region.

8.5.12.2 Procedure

It is assumed that the fibre, matrix, lay-up, stacking sequence, thickness and design load level (P_{Des}) have been determined.

Single or two-row joints

1   No restrictions on dimensions:
(a) select a fastener diameter to give d/t in the range 1 to 3;
(b) select e and w corresponding to the 'knee' on the curve for the appropriate material (Fig. 8.20–8.23);
(c) read off value of bearing strength σ_b;
(d) calculate fastener load from:

$$P_f = \sigma_b dt \, (k_1 k_2 k_3 k_4 k_5 k_6)$$

where k_1 is a correction for d/t (Fig. 8.24);
k_2 is a correction for load direction (Section 8.5.11.1);
k_3 is a correction for environment (Section 8.5.11.3);
k_4 is a correction for configuration (Section 8.5.11.6);
k_5 is a correction for single lap joints (Section 8.5.11.4);
k_6 is a correction for hole elongation (Section 8.5.11.7);
$t = t_p$ if countersunk fasteners used (Section 8.5.11.5);
(e) if $P_f < P_{Des}$, select larger fastener and repeat calculation;
(f) if $P_f < P_{Des}$, and sufficient width is available, calculate number of fasteners in a row to carry P_{Des};
(g) correct for hole proximity using K from Fig. 8.25, giving:

$$P_{joint} = nKP_f$$

(h) repeat for larger fasteners if $P_{joint} < P_{Des}$.
2   Width restriction (limiting number of fasteners in a row):
Following the procedure in (1) it is found that $P_{joint} < P_{Des}$. Employing two rows can raise P_{joint} by 10% (see Section 8.5.11.6). If this is insufficient, add extra plies in joint region until adequate thickness is obtained to give required load. The lay-up may be modified to enhance strength, if appropriate.
3   End distance and/or width restriction:
Repeat above procedure with σ_b corresponding to the appropriate value of e/d (and/or w/d).

Line or multirow joints
If the above procedures have not produced a joint of adequate strength, it will be necessary to consider a more complicated arrangement involving a line, or several rows, of fasteners. A stress analysis procedure that allows for fastener flexibilities and material non-linearity is needed (see Section 8.5.10).

Consideration should also be given to 'tailoring' the laminate. By changing the ply arrangement across the width of a panel fasteners can be placed in a different lay-up to the regions between fasteners.[8.17]

8.6 Bolted–bonded joints

Bolting and bonding together do not confer any significant advantage over a well-designed bonded joint in an undamaged structure.[8.18]

There is a case for using rivets as tooling aids when bonding, since these then act as anti-peel devices.

Bolts can significantly enhance the strength of a damaged bonded joint.

8.7 References

[8.1] *Engineered Materials Handbook*, Vol. 1 *Composites*, Ohio, ASM International, 1988.

[8.2] J. Shields, *Adhesive Handbook*, (3rd Edn), Guildford, Surrey, Butterworths, 1985.

[8.3] I. Skeist, (Ed.), *Handbook of Adhesives*, New Jersey, Van Nostrand Reinhold, 1990.

[8.4] L.J. Hart-Smith, Chap. 7 in *Joining of Fibre-Reinforced Plastics*, (F.L. Matthews, (Ed.)), Barking, Elsevier Applied Science, 1987.

[8.5] L.J. Hart-Smith, CR-2218, NASA, 1974.

[8.6] L.J. Hart-Smith, CR-112236, NASA, 1973.

[8.7] K.C. Kairouz, PhD Thesis, Imperial College, University of London, March 1991.

[8.8] L.J. Hart-Smith, CR-112235, NASA, 1973.

[8.9] L.J. Hart-Smith, CR-112237, NASA, 1973.

[8.10] T.A. Collings, Chap. 2 in *Joining of Fibre-reinforced Plastics* (F.L. Matthews, (Ed.)), Barking, Elsevier Applied Science, 1987.

[8.11] L.J. Hart-Smith, Chap. 6 in *Joining of Fibre-reinforced plastics* (F.L. Matthews, (Ed.)), Barking, Elsevier Applied Science, 1987.

[8.12] CP – 427, AGARD, 1988.

[8.13] G. Kretsis and F.L. Matthews, *Composites*, April 1985, **16**, 2.

[8.14] J.M. Hodgkinson, D.L. de Beer and F.L. Matthews. SP-243, European Space Agency, 1986.

[8.15] F.L. Matthews, E.W. Godwin and P.F. Kilty. TN 82-105, Dept of Aeronautics, Imperial College of Science, Technology and Medicine, London, 1982.

[8.16] T.A. Collings, *Composites*, July 1982, **13** 3.

[8.17] J.R. Eisenmann and J.L. Leonhardt, in *Joining of Composite Materials* (K.T. Kedward (Ed.)), ASTM STP 749, 1981.

[8.18] L.J. Hart-Smith, *J. Aircraft*, November 1985, **22** 11.

Part V

Case studies

The following case studies have been incorporated into this book to illustrate the long term behaviour of various composite structures built over the last two or three decades. Many examples of composite structures have been erected recently, but as yet they have not been exposed to the natural environment for a sufficiently long time to allow an opinion to be formed on their in-service behaviour. These recently built structures include the American Express Building in Brighton, built in the early 1980s, and the Aberfeldy cable stay polymer composite bridge over the river Tay built in 1992.

The case studies given are:

Chapter 9: the design of a GFRP marine vessel.

Chapter 10: the walkway at Terminal 2, Heathrow Airport, which has been in existence for 18 years. The discussion here refers to the fire performance of the polymer composite material.

Chapter 11: the design of a commercial vehicle sideguard system. There were no reported problems regarding the fixings of the system to conventional materials but the author emphasises the possible problems in the actual utilisation of these guards.

Chapter 12: the design and production of a composite component for the automotive industry is discussed and the data requirements for producing a large GFRP structural component are illustrated. The article highlights the development procedures required by the industry.

Chapter 13: the chapter discusses the use of pultruded profiles in a structural application and gives an example to illustrate the design procedure.

Chapter 14: the preproduction test structure, for a GFRP mine-sweeper, was manufactured from a box core sandwich construction. The design requirements and approach are discussed. These structures were first manufactured in the mid-1960s and have consequently yielded considerable information regarding the long term behaviour of composites.

Chapter 15: the case history discusses the shield, made from a sandwich structure of polyester impregnated glass fibre woven roving face material and an end grain balsa wood core material; the shield was for Vickers' 4.5 inch mark 8 gun. Again this structural system was manufactured in the mid-1960s.

9 Design of bottom shell laminate for GFRP vessel

Glass fibre reinforced polymers (GFRPs) have been used extensively for the production of small boats since the introduction of these materials to the marine industry, now some 40 years ago. It is now the accepted material for all production boats, together with many one-off boats and it has a growing use in fast ferries, larger vessels such as the MCMV Class built by Vosper Thornycroft Ltd and other displacement vessels. Furthermore, GFRP has a growing market in the offshore industries where it is being used for lower weight structures in blast protection, pipes, pressure vessels, etc., and for submersibles and sub-sea structures.

The introduction of the higher strength fibres, such as aramid and carbon, has enabled the industry to exploit further the lower weight of reinforced polymers for those vessels seeking optimum performance – racing yachts, power boats, surface effect ships, etc.

The marine industry has been a driving force in the development of these polymeric materials owing to the very aggressive environment in which the structures have to perform. Consequently there are now many acceptable polyester resin systems, vinyl esters, epoxies and purpose-designed resin systems.

Reviewing the role played by reinforced polymers in the marine industry, the main advantages and disadvantages of the materials can be summarised as follows:

1 Advantages:
  (a) excellent environmental performance (no corrosion);
  (b) excellent formability;
  (c) high specific strength and stiffness properties leading to low weight;
  (d) good surface finish;
  (e) large range of fibres leading to ease of structural optimisation by selection of fibre type and orientation;
  (f) good fatigue properties;
  (g) low thermal conductivity;
  (h) non-magnetic.
2 Disadvantages
  (a) high material cost of the high performance fibres;

(b) combustibility of some resin systems;
(c) the need for good quality control during manufacture;
(d) the cost of the necessary production tooling;
(e) lack of comprehensive design standards and some material data.

Probably the greatest inhibitor for the use of fibre reinforced polymers (FRP) for the larger vessels and structures is the tooling cost. It is perhaps a little surprising that the industry has not developed alternative processing methods to contact moulding, which requires a male or female tool. Consequently, the tooling needs to be amortised over the units produced and if these are low in number or indeed one-off, then FRP cannot compete with welded aluminium alloy, despite its relatively lower weight. For production boats, though, FRP stands alone as the most effective material, producing low weight structures at minimum cost.

The type of structure will determine the design. For instance, performance sailing yachts seek weight-effective stiff structure which leads to the use of sandwich construction, incorporating cores such as PVC foam and end grain balsa. Where minimum weight is sought, honeycomb cores are used, such as Nomex (Du Pont's registered trade mark).

Fast vessels, such as racing power boats and high speed passenger ferries, are subjected to water slamming pressures and therefore require an energy absorbing structure below the water line. This can be best achieved by using a structure of single skin stiffened internally by ribs, frames and bulkheads. Such a structure has a degree of resilience and can therefore absorb energy at the water interface. These vessels will, however, invariably use sandwich structure for the decks and superstructures as it provides the lightest structure, although the costs of core materials need to be carefully investigated to avoid a cost penalty compared with single skin structures. Sandwich structure is also used for the frames, bulkheads and other internal supporting structure.

The overall design of a marine structure in FRP, or indeed any FRP structure for that matter, is a carefully balanced integration of parameters, needs and materials, utilising the skills of several professions – structural design, stress analysis, processing, buying, naval architecture, interior design, etc. There is no set solution for any product; each must be carefully worked through to achieve the optimum result with respect to weight and cost. The latter must include through life cost, because reduced weight and improved environmental performance has a knock-on effect through the life of the product.

It is most important that at the conceptual stage of the project the structural design is an integrated function of the overall design. Retrospective design, of any sort, only leads to unacceptable compromises.

The first job of the structural designer is to decide on the structural format and basic materials and to liaise with those responsible for production as to how the vessel is to be constructed. All three factors have a direct effect on the resulting production cost. It is also important to decide on the policy of overall weight and whether the vessel is to be built to any classification requirements.

The vessel weight can be reduced by:

1 Structural optimisation by calculation.
2 Selecting higher performance materials.
3 Structural format, i.e. sandwich or single skin, longitudinally or transversely stiffened.

The type of vessel will generally determine the degree to which weight reduction is taken. Performance vessels require the lowest weight without restraints of cost. Production vessels, particularly powered vessels, still benefit from minimum weight since speed can be increased for the same power, but generally the cost of such vessels does not permit the use of the higher cost high performance fibres and resin systems, i.e. carbon–epoxy prepreg materials (epoxy pre-impregnated fibres requiring heat cure). Suffice it to say, any excess weight should be avoided as this represents money, and all producers have an interest in keeping the build costs to a minimum.

The Classification Societies have Rules for scantlings laid down for different vessel types. These Rules are empirical and are based on many years of experience with respect to the performance of such vessels. However, being empirical and only covering the primary scantlings, they do not lead to minimum weight structures. This can only be achieved by the use of first principle calculations. Should Classification approval be required, then this can be obtained by submitting the structural calculations along with the drawings to the chosen Classification Society.

Experienced designers can make the basic decisions on structural format and materials relatively quickly for any new vessel as they will have a data bank of information from previous designs. Those starting from scratch and new to the industry will be forced to explore many options before concluding on the basic points, such as whether the vessel should be longitudinally or transversely stiffened and where and when not to use sandwich structure to the benefit of weight and cost.

The selection of materials can also be a complex procedure as there are so many different fibres and materials, besides different weaves of fibres, finishes, foam cores, etc. The basic constituent of GRP as used in the marine industry are 'E' glass fibres and polyester resin, with end grain balsa wood and PVC foam as core materials for sandwich structure. Gel coats are favoured to provide a good surface finish and they should be selected with care, ensuring that they have adequate strain compatibility with the laminating resin and that they have acceptable environmental performance. To avoid blistering, it is essential that an

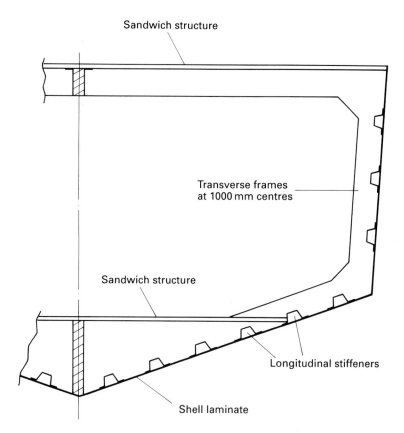

Sandwich structure

Transverse frames
at 1000 mm centres

Sandwich structure

Longitudinal stiffeners

Shell laminate

9.1 Typical midship section.

isophthalic gel coat is specified if polyester resins are to be used, together with isophthalic resins for the laminate. Should higher performance be required then aramid or carbon fibres are specified, either on their own or as hybrids. A hybrid could either be a mixed weave of different fibres or a laminate with plies of different fibres. This highlights one of the advantages of FRP, the ability to introduce discreet layers or bundles of fibres to increase strength or stiffness locally.

In order to illustrate the basic concept of design, a particular vessel will be considered as an example. Typical for the larger pleasure boat market is a 20 metre vessel of Vee bottom configuration. Maximum speed in calm water would be around 30 knots and 25 knots in sea state 3. A midship section of the hull is shown in Fig. 9.1.

The overall structural design is an extensive operation if weight and cost are to be optimised. The complete design of the vessel is beyond the scope of this example, so two areas will be selected to highlight the design philosophy and types of structure. The bottom shell laminate and stiffening will be worked through together with the deck sandwich structure.

It will be assumed that a standard build configuration has been decided, using a conventional female mould and a contact moulding manufacturing process. It will also be assumed that 'E' glass fibres and polyester resin have been selected as the principal fibre–resin system.

As this is a vessel which will be affected by wave slamming, then a single skin bottom and topsides construction will be considered with a transverse framing system and longitudinal skin stiffeners. Sandwich structure will be used for the deck. This represents a fairly classic solution to the structural format and material selection.

9.1 Bottom skin/stiffener design

The predominant loads on this area of structure are wave slamming and hydrostatic. The calculation of wave slamming forces has received more attention in recent years and empirical methods are available to derive the forces for particular vessel types.[9.1, 9.2] Alternatively, the designer can turn to Classification Rules which, in some instances, provide similar empirical methods for calculating forces for certain types of vessels.[9.3, 9.4]

For this particular vessel, the slamming pressure at the midship section has been calculated to be 180 kPa for the shell skin and stiffeners and 150 kPa for the supporting structure. Similar figures of 120 kPa and 80 kPa apply to the topsides.

The hydrostatic forces can be calculated for the assumed head of water. Note that if Classification acceptance is required then such hydrostatic pressures may be determined by a given head for a particular vessel and all forces should be agreed with the particular Classification Society before detailed calculations are undertaken.

The selection of factors of safety will be governed by Classification, if this is to be achieved. The various Classification Societies will have definite factors of safety for different loadings and areas of structure. Typically, though, 1.50 is used for slam pressure and 3.0 to 4.0 for hydrostatic pressure. For deck loads and internal masses (engines, gearboxes, etc.) a figure of 3.0 to 4.0 is common. In addition to this, it is usual to limit the deflection of decks and superstructure to 1% of the span.

The selection of material design properties is best undertaken against a valid set of data derived by testing equivalent materials. The designer should not rely on manufacturer's data as this may have been determined using unobtainable quality control and environmental conditions. Where new materials are being used, then eventual testing is the only way to ensure reliable data.

A sound data bank will permit the designer to select laminates incorporating plies of different materials by using suitable software,[9.5] but again, it must be emphasised that such 'engineered' laminates will require validation of properties by testing laminates made under compatible conditions to those of the chosen manufacturing procedure and the selected environment.

Care must also be exercised with respect to fatigue life and creep rupture where repeated loadings are dominant or where the application of a long term steady load needs to be accommodated.

The bottom shell is to be designed for the slam pressure such that the selected factor of safety is achieved. To minimise the weight of the shell laminate, it is beneficial to allow the skin to deflect so that a degree of tensile membrane stress is developed. The optimum thickness of the shell laminate is a function of the pitch of the internal stiffening, but is also governed by the selection of a minimum practical thickness. Such a minimum is obviously necessary as a very thin shell, although perhaps having adequate strength, could be punctured by occasional impact with floating debris. The Classification Rules lay down such minimums and they provide a useful guide.

The minimum shell thickness for this vessel has been established using Classification Rules as a basis and calculated to be 6.0 mm for the selected material for 'E' glass woven roving reinforced polyester resin. This provides the starting thickness for pitching the stiffeners. It must be noted, however, that the minimum weight is achieved by having the minimum shell thickness and the closest stiffener pitch. The cost of manufacture, though, will be reduced if the number of internal stiffeners is reduced, requiring a thicker shell thickness. This is a balance which the experienced designer can judge, and the ratio of hull thickness to stiffener spacing can be arrived at by previous designs, data and experience.

Likewise with the selection of the weights of reinforcing materials. Some vessels require a very high degree of surface finish without any 'print-through' induced by woven fabrics. The common method of producing such a good finish is to use chopped strand mat behind the gel coat but this reduces the overall properties of the laminate. It is better, from a structural point of view, to use a lighter weight satin weave woven roving laminate behind the gel coat, which will reduce or even eliminate print-through, if care is taken.

For this design exercise, it will be assumed that a standard weight woven roving of 800 g/m^2 will be used at a resin : fibre ratio of 1.0 : 1.0, i.e. a weight fraction of 0.50. The thickness of each layer is calculated to be 0.98 mm. The layers adjacent to the gel coat could be replaced with lighter material to assist surface finishing.

The calculation of shell laminate thickness under large deflections (greater than half laminate thickness) can be conveniently evaluated by plotting tensile stress versus panel span. Use can be made of the

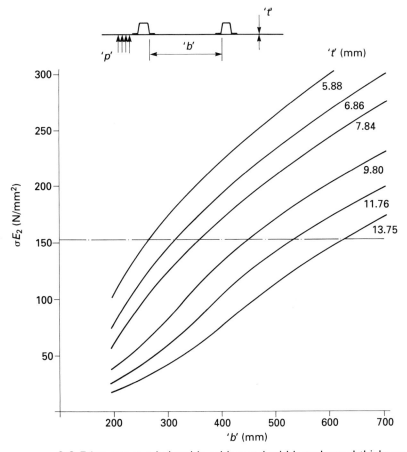

9.2 Edge stress relationship with panel width and panel thickness.

Engineering Science Data Unit (ESDU) data sheets,[9.6] which, although for isotropic materials, can be used with sufficient accuracy for a balanced weave such as woven roving and where the reinforcement is aligned to the panel edges. Figure 9.2 provides the curves for various thicknesses based on increasing plies of 800 g/m² woven roving, calculated for a pressure of 180 kPa and varying values of panel width. These curves can now be used to select a skin thickness and panel width, bearing in mind a practical minimum of 6.0 mm.

Typical material properties for a plain weave woven roving aligned at 0°/90° and with a weight fraction of 0.50 can be considered as:

Tensile stress: 228 N/mm²
Tensile modulus: 13 800 N/mm²
Compression stress: 186 N/mm²

Compression modulus: 13 800 N/mm²
Shear strength: 70 N/mm²
Shear modulus: 3000 N/mm²
Interlaminar shear strength: 14 N/mm²
Weight fraction: 0.50.

To achieve a factor of safety of 1.5 on tensile strength, a limit of 152 N/mm² must be set on the curves.

A range of options are available for skin thickness versus stiffener span. As previously stated, the thinner the shell laminate, the lower the total structural weight, but the cost will increase because of the greater work involved from the increased number of stiffeners. For this vessel a practical choice is a laminate thickness of 7.84 mm (eight plies of 800 g/m² woven roving) and a panel span of 350 mm, a balance between cost, weight and robustness.

The curves in Fig. 9.2 plot the maximum edge tensile stress. The ESDU data sheets will also provide the membrane tensile stress which enables the corresponding inner face compression stress to be calculated, together with data for the stresses at the plate centre. These should also be checked and the margins on strength established.

The use of the ESDU data sheets does not take into account the thickening at the panel periphery caused by the stiffener and frame overlays. The stress and deflection values can be obtained by the use of mathematical modelling and solving using finite element analysis. The effect of the edge stiffening is to reduce the stresses at the edges and at the centre. However, such analysis is reasonably lengthy and unless a great deal of weight optimisation is sought, these procedures are not normally used. If they are used, then the corresponding degree of care is needed during manufacture to ensure that all tolerances on both thickness and geometry are not exceeded. Confidence is also required as to the accuracy of the applied loads, in this case, wave slamming forces. For production vessels it is perhaps prudent to use the data sheet approach with the knowledge that an additional margin on strength is achieved by the edge thickening.

The bottom shell laminate stiffener can now be investigated. The design of the stiffener depends on the frame pitch, which is partly governed by the internal geometry. Generally speaking, a pitch of one metre is typical for such a vessel and will be considered for this exercise. Also, typically stiffeners are 'hat' shaped, having a base dimension of about 150 mm and tapered to ease laminating. The stiffener is therefore to be designed for a load of $180 \times 10^{-3} \times 500 = 90$ N/mm.

It is beneficial to incorporate unidirectional reinforcement to the crown of the stiffener, which helps to pull the neutral axis of the section more towards the middle whilst at the same time increases the overall load-carrying ability.

Figure 9.3 provides the details of the selected stiffener section which

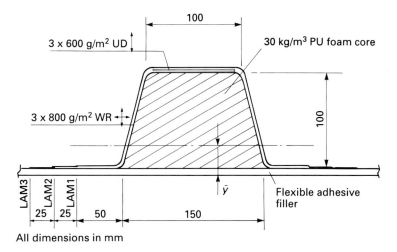

9.3 Shell laminate stiffener.

incorporates 800 g/mm² woven roving material over a nonstructural pre-form core. The section modulus has been calculated allowing for the higher modulus of the unidirectional reinforcement (25 000 N/mm²):

Moment of inertia$=7371\times10^3$ mm⁴ (relative to the woven roving)
$\bar{y} = 27.26$ mm

The stiffener is continuous, so a maximum bending moment at either the support or the mid-span will be approximated to:

$$BM = 0.10\ wl^2 = 0.10 \times 90 \times 1000^2$$
$$= 9.0 \times 10^6\ \text{N mm}$$

The maximum tensile stress in the unidirectional reinforcement is developed at the stiffener mid-span, and is calculated using the relative modulus of elasticity values:

$$\sigma_t = \frac{9.0 \times 10^6\ (112.76 - 27.26)}{7371 \times 10^3} \times \frac{25\ 000}{13\ 800}$$

$$= 189\ \text{N/mm}^2$$

The maximum compression stress in the unidirectional reinforcement is at the stiffener support and has the same value as the above compression stress.

Typical properties for unidirectional reinforcement are:

$\sigma_t = 500$ N/mm²
$\sigma_c = 400$ N/mm²
$E_t = E_c = 25\ 000$ N/mm²
Weight fraction $= 0.54$

To achieve the selected factor of safety of (1.50), these values are reduced to 330 N/mm² (tension) and 276 N/mm² (compression), which when compared to the calculated applied stresses are satisfactory.

Likewise the maximum stresses in the woven roving can be calculated:

$$\sigma_t \; = \; \sigma_c \; = \; \frac{9.0 \times 10^6 \times 85.5}{7371 \times 10^3}$$

$$= 104 \text{ N/mm}^2$$

The permissible stresses are 150 N/mm² (tension) and 120 N/mm² (compression). The stiffener is therefore satisfactory for the bending stresses.

The maximum shear force in the stiffener is at the frame interface or support:

$$S \; = \; \frac{90 \, (1000 - 50)}{2}$$

$$S \; = \; 42.8 \text{ kN}$$

assuming the frame is 50 mm in width.

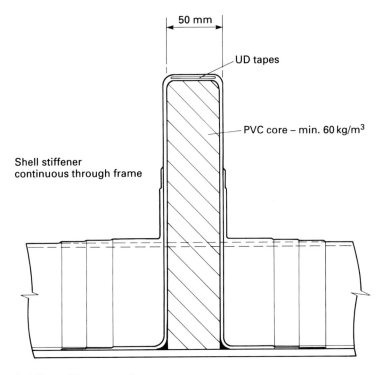

9.4 Typical frame section.

Therefore:

$$\text{shear stress } \sigma_s = \frac{42.8 \times 10^3}{100 \times 2 \times 2.94} = 73 \text{ N/mm}^2$$

This exceeds the permissible value of 0.67×70 N/mm², and therefore the thickness either needs to be increased or the web material turned through to $\pm 45°$. This latter method, however, will reduce the bending strength. The thickness is therefore increased to five plies of 800 g/mm² ($t = 4.9$ mm), reducing the shear stress to 44 N/mm². This thickness will apply only to the ends of the stiffener; 200 mm from the frame.

The framing is designed for the appropriate loads, bending moments and shear forces. A typical frame section is shown in Fig. 9.4. It is normal to use a structural core in frames, such as PVC, to give adequate stability to the frame sides.

9.2 Deck Sandwich Structure

The decks are usually designed in a sandwich structure as this is the lightest and cheapest way of achieving adequate stiffness and strength. Deck loads vary as to the location and use. For this example, a typical value of 20 kPa will be used.

The deck sandwich considered in this design example will be taken as 'E' glass woven roving at 800 g/m² per ply over a core of PVC foam, this being typical for many such structures.

The sandwich structure shown in Fig. 9.5 will be checked for strength and stiffness, together with local stability. There are various methods published on the design of sandwich panels, which take into consideration the sandwich constituent properties and configuration, i.e. isotropic or orthotropic cores (PVC as opposed to, say, honeycomb). For this design example, the sandwich will be checked using various charts which are reproduced for reference.

The panel has a width defined by the frame spacing, i.e. 1000 mm, and it will be assumed that the panel aspect ratio is 0.30 with a uniform load and having simply supported edges.

Notation used:

E_f : face laminate modulus of elasticity;
E_c : core modulus of elasticity;
G_c : core shear modulus;
t : face laminate thickness;
t_c : core thickness;
d : overall sandwich thickness;
b : panel width;
v : Poisson ratio of face laminate.

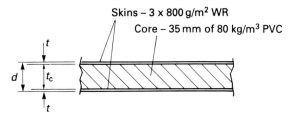

9.5 Deck sandwich structure.

Calculate the stiffness parameter:

$$V = \frac{\pi^2 E_f t t_c}{2\lambda b^2 G_c}$$

where $\lambda = (1 - v^2)$.

For 80 kg/m³ PVC core $G_c = 31$ N/mm² and:

$$V = \frac{\pi^2 \times 13\ 800 \times 2.94 \times 35}{2 \times 0.98 \times 1000^2 \times 31}$$

$$= 0.231$$

The maximum bending moment will occur at the panel centre, and the moment across the width:

$$M = \frac{16pb^2}{\pi^4} (C_3 + vC_2)$$

where p = deck load of 20 kPa
C_2 and C_3 are obtained from Fig. 9.6 and:

$$M = \frac{16 \times 20 \times 10^{-3} \times 1000^2}{\pi^4} (0.03 + 0.13 \times 0.72)$$

$$= 0.41 \text{ kN/mm}$$

$$\text{Facing stress} = \frac{2M}{t\ (d + t_c)}$$

$$= \frac{2 \times 0.40 \times 10^3}{2.94\ (40.88 + 35)}$$

$$= 3.59 \text{ N/mm}^2$$

The maximum shear load (on long side):

$$S = \frac{16pb}{\pi^3} C_4$$

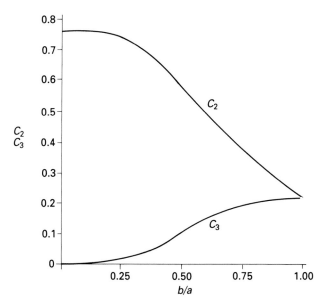

9.6 Values of coefficients C_2 and C_3 for various values of ratio b/a and isotropic core material.

C_4 is obtained from Fig 9.7 and:

$$S = \frac{16 \times 20 \times 10^{-3} \times 1000 \times 0.97}{\pi^3}$$

$$= 10.0 \text{ N/mm}$$

and:

$$\sigma_s = \frac{2S}{d+t_c}$$

$$= \frac{2 \times 10.0}{(40.88 + 35)}$$

$$= 0.263 \text{ N/mm}^2$$

The shear strength for 80 kg/m^3 PVC foam = 1.0 N/mm^2. This does not quite achieve the required factor of safety of 4.0 but is considered acceptable as no account is taken for the shear strength contribution from the facing skins.

$$\text{Panel deflection } \delta = \frac{16pb^4}{\pi^6 D} C_1$$

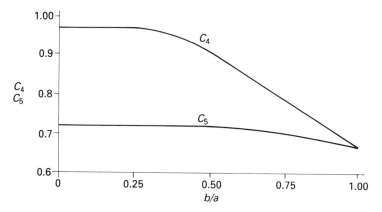

9.7 Values of coefficients C_4 and C_5 for various values of ratio b/a and isotropic core material.

where:

$$D = \frac{E_f t h^2}{2\lambda}$$

for equal facings, where $h = t_c + t$:

$$D = \frac{13\ 800 \times 2.94 \times 37.94^2}{2 \times 0.98}$$

$$D = 29.8 \times 10^6 \text{ N mm}$$

C_1 is obtained from Fig. 9.8 and:

$$\delta = \frac{16 \times 0.02 \times 1000^4}{\pi^6 \times 29.8 \times 10^6} \times 0.90$$

$$= 10.0 \text{ mm}$$

Allowable deflection is 1% of span = 10.0 mm.

The sandwich structure in this case is therefore deflection and shear stress limited. However, it is also prudent to check the local face in-stability (wrinkling) stress by the following method. Determine parameter q:

$$q = \frac{t_c}{t}\ \frac{G_c}{}\left(\frac{\lambda}{E_f E_c G_c}\right)^{1/3}$$

$$= \frac{35 \times 31}{2.94}\left(\frac{0.98}{13\ 800 \times 85 \times 31}\right)^{1/3}$$

$$= 1.11$$

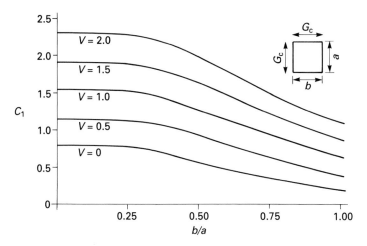

9.8 Values of coefficients C_1 for various values of ratio b/a and isotropic core material.

Determine parameter K:

$$K = \frac{E_c \eta}{F_c t_c}$$

η is the total amplitude of the facing waviness, which will depend on the core type and density. For design purposes, assume a value of 5% of the facing skin thickness. Then:

$$K = \frac{80 \times 0.015}{1.2 \times 35} = 0.029$$

The wrinkling stress:

$$F_{cr} = Q \left(\frac{E_f E_c G_c}{\lambda} \right)^{1/3}$$

Q is obtained from Fig. 9.9 and:

$$F_{cr} = 0.46 \left(\frac{13\,800 \times 85 \times 31}{0.98} \right)^{1/3}$$

$$= 153 \text{ N/mm}^2$$

Permissible $F_{cr} = 0.25 \times 153$
$$= 38 \text{ N/mm}^2$$

This is to be compared with the permissible face compression stress of $0.25 \times 186 = 46.5 \text{ N/mm}^2$ and the applied stress of 3.65 N/mm^2.

It can be seen from the above example that there are numerous permutations of materials, geometry and sandwich panel configurations

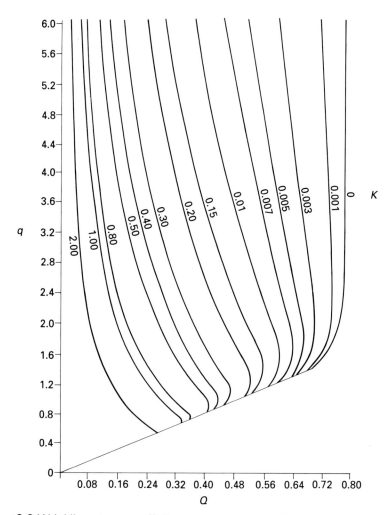

9.9 Wrinkling stress coefficients.

which require optimisation, should weight and cost be fully exploited.

As an indication of the effect of facing skin material on weight, Fig. 9.10 plots the weight per square metre of PVC core and Nomex core sandwich deck structure for different facing skin reinforcement under a constant applied load normal to the surface.

By moving the 'E' glass–polyester contact moulded structure to carbon-epoxy prepreg structure, a 30% weight saving can be achieved, or 38% if Nomex is used. However, the associated manufacturing cost increases over four times that of the basic 'E' glass–polyester structure. The weight decrease and cost increase need to be quantified with respect to overall performance and through-life costs.

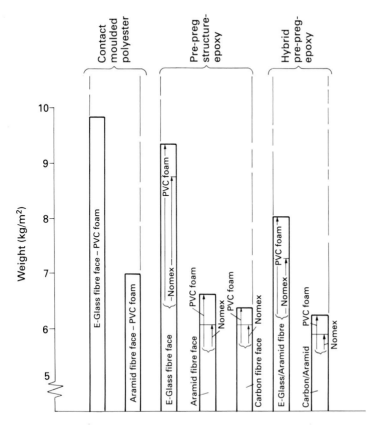

9.10 Comparative weights of sandwich structure with varying cores and skin reinforcing–resin systems.

9.3 References

[9.1] D. Savitsky and P. Ward Brown, 'Procedures for hydrodynamic evaluation of planing hulls in smooth and rough water', *Marine Technology* 1976 **13** (4) 381–400.

[9.2] R.G. Allen and R.R. Jones, 'A simplified method of determining structural design – limit pressures on high performance marine vehicles', AIAA/SNAME Advanced Marine Vehicles Conference, 1978.

[9.3] Det Norske Veritas Classification A/S Tentative Rules for Classification – High Speed and Light Craft, 1991.

[9.4] Lloyds Register of Shipping, Provisional Rules for the Classification of High Speed Catamarans, 1991.

[9.5] Think Composites Software: Tsai 1992.

[9.6] Engineering Sciences Data Unit, Item No. 71013, 'Elastic direct stresses and deflections for flat rectangular plates under uniformly distributed pressures'.

10 The walkways at Terminal 2, Heathrow Airport: Designing for fire performance

10.1 Introduction

In the mid-1970s Terminal 2 at Heathrow was being extensively redeveloped. One of the objectives was to give passenger arrivals on the first floor access to ground level without mingling with passengers checking in on the ground floor, which was becoming increasingly crowded. For this reason two external walkways were planned leading from first to ground level. Terminal 2 was a prize-winning design and had a brick facade, and it was proving difficult to come up with a design of walkway which blended in with the original design. It was therefore decided to come up with a contrasting design which was obviously a subsequent addition to the original building.

At that time glass reinforced plastics (GRP) were making great advances and some new buildings were being distinctively clad with GRP panels. It was therefore decided to try to design the walkways in GRP.

The architect's brief required direct passenger access between the two main levels, the bridge to the car park and the newly opened Underground Station. It was acknowledged that users would be pushing baggage trolleys through the walkways. It was also evident that since there would be limited space available on site during construction, a solution which incorporated lightweight factory-made units to be assembled on site would be optimum. It was believed that the structure should have simple and clean lines derived from modern technology. As a result of these considerations a GRP tube was thought to be the answer.

When working with GRP the smallest possible number of different panels is desirable since the cost of moulds is high and the number of panels made from each mould is likely to be low. At first it was expected that 4 different panels would be needed, but 24 were eventually required. Since a fair face was needed on both the inside and the outside of the walkway the panels needed to be double skinned, necessitating two moulds to each type of panel.

The architect's objective was to maximise the use of GRP and to minimise the use of steel. The steel frame consisted of 400 × 200 mm mild steel box sections, positioned at the waist of the tube and inclined at the gradient of the ramp. They were supported either on columns at 3 m intervals at the departure end and on the concrete structure at the arrivals end. An open grill steel tray formed the basic floor slab. The upper GRP panels sit on the steel box sections and are designed to have adequate structural strength and stiffness. The internal dimensions of the walkways are height 2.7 m, width at base 2.5 m, width at waist 3 m and width at the top 2.5 m. The span of the structure, i.e. 3 m, is relatively modest but it opens up at the entrances to the main building into wide foyers, the one at the departure end being 9 m wide. The structure incorporated a number of large windows approximately 1 m high by 0.75 m wide all set in GRP frames. There are windows in every panel at the return bends and others on the straight lengths.

10.2 Fire performance

The local authority fire office designated the walkways as escape routes. This meant that in the case of an external fire they should maintain their integrity for sufficient time for the occupants to escape in safety and that the internal surfaces should not spread flame or contribute to the fire.

In technical terms the specification was:

1 An inside finish to meet Class 0.
2 An outside finish to meet Class 1 when tested to BS 476: Part 7 (Surface Spread of Flame).
3 The walkways, ramps and their associated steel beam supports and decking should be structurally half-hour fire-rated against an external fire to BS 476: Part 8 and that the windows should meet this rating as well.

At that time a Class 0 rating under the UK Building Regulations (Class 0 is not a British Standard classification) required that the material should not catch alight when subjected to a small flame and should have a fire propagation index when tested to BS 476: Part 6 of less than 12 after 20 minutes.

At present Class 0 means a Class 1 surface spread of flame when tested to BS 476: Part 7 and an index i_1, after 3 minutes, of not greater than 6 and a total (Final) index (I) of not greater than 12 after 20 minutes when tested to BS 476: Part 6.

The fire propagation test is a relatively crude method of measuring the heat released by a burning sample and its contribution to the rate of temperature rise and hence the speed of development of the fire. The surface spread of flame tests measured the rate at which flame will spread along or up a surface: Class 1 is the best classification.

A half-hour fire resistance means in broad terms that if there is a fire on the outside of the walkway then the walkway will not collapse, or crack to let flames penetrate for 30 minutes and for the same period the temperature on the side not in contact with the fire must not rise by more than 180 °C so that flammable materials against the wall on the inside will not catch fire. BS 476: Part 8 is now obsolete but the test was essentially the same as described in the present BS 476: Parts 20 to 23.

10.3 Theory of fire performance

Fire is a complex subject. No two fires are exactly the same. The fire's rate of development and its intensity depend on the materials involved and the amount of oxygen available to support combustion. Materials that give out heat quickly accelerate fire development, while materials that burn across their surface assist the fire to spread.

Many materials which are at first sight non-combustible will under some conditions ignite. Steel is an example. In the form of steel wool it will rapidly oxidise when a flame is applied and it catches fire. In a developing fire where there is a great deal of radiated heat, materials which will not normally catch fire when a small flame is applied; under fierce radiant heat will increase in temperature and spontaneously ignite, spreading the fire. This is call flashover.

Faced with this complexity, those framing regulations have to try to use simple criteria to control and reduce the risk. The easiest solution would be to specify that only materials that cannot burn should be used but that would make our lives quite unbearable. For instance, wood would be banned as would most cloth and all plastics foam used for upholstery and for thermal insulation and all polymeric materials. Polymeric materials are organic – they are principally formed from carbon, hydrogen and nitrogen and hence they are flammable to various degrees. In spite of this, they have increasingly penetrated the building and construction market although undoubtedly their progress has been slower because of their poor flammability performance.

There is a myth lingering in the plastics industry that Building Regulations have been framed as a barrier to prevent plastics being used. This is definitely not the case. Fire regulations have been in force for hundreds of years and the present philosophies have been in existence since the 1930s, long before plastics were being considered for building use. The regulations were framed around the level of risk offered by traditional materials such as wood, horse hair, stone, concrete and bricks. Undoubtedly, our whole concepts of acceptable flammability fire spread and safety have been framed around these materials and this has led to a whole culture of the use and control of flame sources in our lives. Increasing the risk by the introduction of more flammable materials is difficult but when the advantages can be clearly seen, as in the case of foam furniture, then they can be accepted. However, there is undoubtedly a wish to return to the status quo as shown by the development towards less flammable foams and covering fabrics and the drive to popularise smoke alarms.

Building Regulations attempt to limit fire risks by:

1 Limiting flame spread.
2 Limiting the heat released by materials of construction.
3 Compartmentalising the building with fire-resisting walls.

Section B1 – Means of escape

Requirement
Means of escape **B1.** The building shall be designed and constructed so that there are means of escape in case of fire from the building to a place of safety outside the building capable of being safely and effectively used at material times.

Section B2 – Internal fire spread (linings)

Requirement
Internal fire spread (linings) **B2.** (1) to inhibit the spread of fire within the building, the internal linings shall – (a) resist the spread of flame over their surfaces; and (b) have, if ignited a rate of heat release which is reasonable in the circumstances (2) In this paragraph 'internal linings' mean the materials lining any partition, wall, ceiling or other internal structure.

Section B4 – External fire spread

Requirement
External fire spread **B4.** (1) The external walls of the building shall resist the spread of fire over the walls and from one building to another, having regard to the height, use and position of the building. (2) The roof of the building shall resist the spread of fire over the roof and from one building to another, having regard to the use and position of the building.

10.1 UK Building Regulations: Section B1, B2 and B4.

While the latest Building Regulations do not specify in detail the methods to be used, they do set out in general terms the objectives and go on in Approved Document B to suggest some ways in which these objectives can be achieved. However, many of the suggestions restrict the use of plastics rather than promote them, but with skilful design and material selection GRP structures can be designed to comply in full with the fire safety requirements of the Regulations. Compliance is demonstrated by specified Fire Tests.

10.3.1 The (UK) Building Regulations 1991

The section of the Regulations applicable to Fire Safety are contained in Approved Document B which is divided into five parts as follows, each with its own requirements:

B1   Means of Escape
B2   Internal Fire Spread (Linings)
B3   Internal Fire Spread (Structure)
B4   External Fire Spread
B5   Access and Facilities for the Fire Service

The three sections of relevance to the walkways are sections B1, B2 and B4. The requirements stated in the UK Building Regulations 1991 are reproduced in Fig. 10.1. B1 is relevant because the walkway was designated a 'means of escape'. The relevant clause is the last one 'being safely and effectively used at all *material times*'. It would be expected that the building would be evacuated within 30 minutes, therefore the walkway must withstand fire for half an hour.

Sections B2 and B4 refer to flame spread over the surfaces – characterised by the BS 476: Part 7 surface spread of flame test – and B2 refers to a reasonable rate of heat release characterised by the BS 476: Part 6 fire propagation test.

10.4 Fire tests

The pre-eminent fire test specification for building structures in the UK is BS 476 which has a number of parts, each designed to assess a particular fire performance of a material, surface or structure. The tests define particular properties which are called for at different stages in the development of a fire. These stages will be discussed later.

The tests are briefly described below. The requirements of the UK Building Regulations are usually complied with by successfully testing to the various parts of BS 476, *Fire tests of building materials and structures*. There are several parts to the standard, each dealing with a different aspect of resistance and reaction to fire. The relevant parts of BS 476 will be discussed briefly to give the outline of the type of properties which it examines.

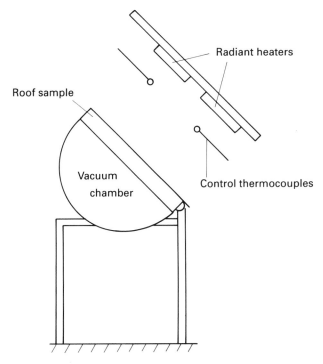

10.2 Roof test apparatus.

10.4.1 BS 476: Part 3 External fire exposure roof test

This test is to determine the performance of a roof and is meant to simulate the effects on a roof of a nearby building catching fire. This will, of course, heat the roof by irradiation and could either cause it spontaneously to catch fire or to give off flammable gases. In addition to this, sparks or flaming brands could be deposited on the roof as a source of ignition.

This situation is simulated by irradiating the roof sample approximately 1 m square by radiant gas-heated panels (Fig. 10.2) These are set parallel to the roof at a distance of approximately 600 mm. During the test a small gas flame is applied to the surface of the sample in order to simulate flaming brands and to ignite any flammable gas which is being given off. Roofs can be tested either horizontal (F) or sloping (S) and this determines the first letter of the classification. The second letter is determined by the time when flame penetration of the sample occurs and the third letter is determined by the distance flames spread across the surface of the specimens.

10.4.2 BS 476: Part 4 Non-combustibility test for materials

This specification includes a definition of non-combustibility and determines whether materials with or without coatings used in construction meet the definition. A cuboid sample with a thermocouple in its centre is lowered into a furnace at 750 °C. The sample shall not flame for more than 10 s or increase the temperature of the furnace or sample thermocouple above the initial furnace temperature by more than 50 °C. In practice, a sample with more than 4% organic content is very unlikely to meet these conditions and so a 'non-combustible' commercially viable plastic or plastic composite is unknown at the time of writing (Fig. 10.3).

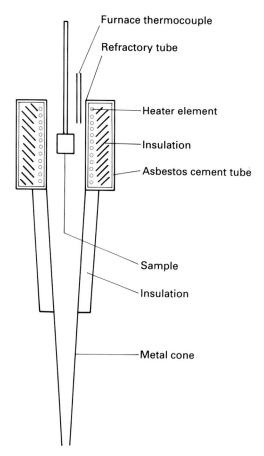

Furnace thermocouple

Refractory tube

Heater element

Insulation

Asbestos cement tube

Sample

Insulation

Metal cone

10.3 Non-combustibility test.

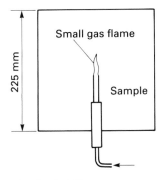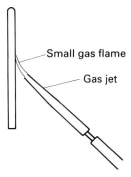

10.4 Ignitability test.

10.4.3 BS 476: Part 5 Method of test for ignitability

This test is solely meant to determine how easily a material will ignite when subjected to a small flame. In the test the flame is approximately 10 mm long and is applied to the middle of a vertical specimen. The fire performance of a material cannot be assessed on this test alone. The practice of polymer manufacturers proudly quoting the result of this test (resistance to ignition by small flaming, thereby implying that this defines adequate fire performance) is to be deprecated. This test refers only to the very start of a fire and not to the material performance when the fire catches hold.

The test rig is shown in Fig. 10.4. It consists of a gas jet which applies a controlled flame approximately 10 mm long to the centre of a vertically held sample sheet 225 mm square, with the flame impinging at an angle of 45°.

10.4.4 BS 476: Part 6 Method of test for fire propagation for products

This test takes account of the combined effects of ignition characteristics, the rate of release of heat and the thermal properties of the products. It is specifically aimed at determining the likelihood of a material to accelerate or contribute to the increasing intensity of a fire. It measures the heat a sample gives out when subjected to radiant heat from electric heaters and small flames from a row of gas jets. The sample is 228 mm square and is one wall of the fire propagation box, as can be seen in Fig. 10.5. The gas jets are ignited at the start of the test and the electric heaters are switched on after 2 min 45 s at 1.8 kW and then altered to 1.5 kW at 5 min. The test lasts 20 min. The progress of the test is monitored by the temperature indicated by thermocouples set in the chimney. Two tests are carried out, one with a non-combustible

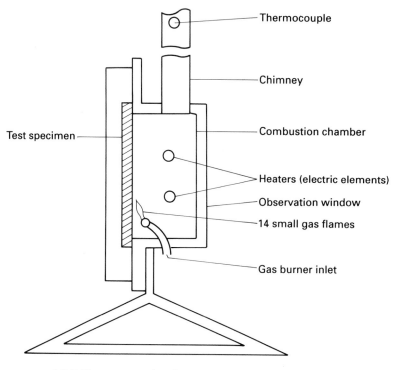

Thermocouple

Chimney

Test specimen

Combustion chamber

Heaters (electric elements)

Observation window

14 small gas flames

Gas burner inlet

10.5 Fire propagation furnace.

sample in place and the second with the test sample. Typical time versus temperature graphs are shown in Fig. 10.6. For a good result the area between the calibration graph and the sample graph should be small. In fact, two indices are calculated, one for the first 3 min and the second for the whole test. These are respectively i_1, which should be not greater than 6, and I which should be not greater than 12. In practice, if the sample burns to any significant extent then heat will be generated and the sample will not comply with the requirements.

10.4.5 BS 476: Part 7 Surface spread of flame test for materials

This test determines the tendency of materials to support spread of flame across their surfaces. It was originally developed to assess whether the spread of flame along the walls of a corridor with a fire at one end was faster than a person could run to escape. It consists of a 1 m² radiant panel which gives out approximately 300 kW. A sample which is 885 mm long by 270 mm high is placed with its short edge approximately 225 mm from the face of the radiant panel, at right angles to it. A small flame approximately 75 mm to 100 mm long plays on the end of the panel nearest the furnace for 1 min. The progress of

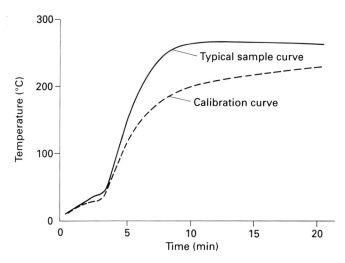

10.6 Fire propagation test curves.

Table 10.1 Flame spread

Classification*	Flame spread at 1.5 min		Final flame spread	
	Limit (mm)	Tolerance for one specimen in sample (mm)	Limit (mm)	Tolerance for one specimen in sample (mm)
Class 1	165	+25	165	+25
Class 2	215	+25	455	+45
Class 3	265	+25	710	+75
Class 4	Exceeding Class 3 limits			

* The flame spread on any specimen of the sample shall not exceed the limit assigned for the class with the proviso that for one specimen only in the sample the flame may exceed this limit by the tolerance shown.

any flame along the sample is monitored and for a Class 1 result the flame must not travel more than 165 mm within the 10 min test duration. A Class 1 result is the best and results up to Class 3 are permitted. These limits can be seen in Table 10.1.

10.4.6 BS 476: Part 8* Test methods and criteria for the fire resistance of elements of building construction

The tests described previously are called reaction to fire tests and are designed to show how a material reacts when subjected to fire conditions, i.e. does it catch fire, does it spread flame or does it give off heat to feed

* now BS 476: Parts 20–23.

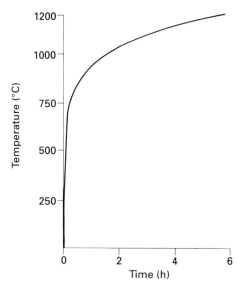

10.7 Fire resistance curve.

the fire? The fire resistance test, on the other hand, is designed to show how a construction will hold back fire. On one side of the large sample, the test both simulates the temperature rise that would occur in a normal fire and gauges the time for which the floor or partition would hold back the fire before it is penetrated or collapses. It is very unusual for a material on its own to be subjected to this test. Usually the material has to be built into a wall or floor with other materials in order to be suitable for the test. The sample is 3 m square and forms the outside wall of a large furnace. The temperature inside the furnace is raised to follow the curve shown in Fig. 10.7. During the test the temperature of the face not exposed to heat is recorded by means of fixed and roving thermocouples. Any gaps that appear are measured. Hot gas and flame penetration or collapse of the sample determine the time to failure, as does an average increase in temperature of 140 °C or any individual thermocouple indicating a temperature increase greater than 180 °C. The reason for the temperature requirement is to make sure that combustible materials on the side of the wall or floor not exposed to the fire will not catch fire and so effectively breach the fire break. Results are given to the nearest minute and structures are usually classified in terms of time to failure designated in 30 minute steps. Thus structures are classified as, e.g., a half-hour partition or a half-hour door, a one hour panel.

10.4.7 The development of a fire

As indicated earlier in this chapter, the different flammability properties of materials are important at different stages in a fire's development. A fire progresses through five stages and the different BS 476 tests can apply to the different stages. The stages are conveniently defined as: ignition; development; flashover; full burning; decay.

In the ignition stage a small flame or hot radiant material comes into contact with a flammable material and causes it to ignite. The BS 476: Part 5 ignitability test obviously refers to this ignition stage, where the fire starts by the material catching fire.

In the development stage the rate of increase of intensity of the fire is governed by a number of factors; the rate of heat release, the thermal inertia and the tendency of the material under consideration to spread flame, the geometry of the fire compartment and fuel load (finely divided material burns more quickly), the fire load of the compartment – usually expressed in MJ/m^2 – and the ventilation factor.

The basic requirement for a fire to develop is that the system as a whole is under a condition of positive feedback. There must be enough energy generated to heat the next piece of material in the fire zone for the reaction to continue. If the energy output is very low then the fire will peter out or simply continue burning slowly but steadily – an open fire in a grate is a good example of this. However, if the energy output is high, and – critically – if the fire is in a small compartment, then the feedback becomes very positive. Under these conditions the developing fire becomes almost a runaway reaction with more heat output involving more material producing more heat output and so on. Finally, the heat and radiation output is so high that all the combustibles in the room become involved, because they all begin to give off flammable smoke which reaches its flashpoint, the temperature rises several hundred degrees centigrade in a few seconds and the whole room becomes engulfed in flame. This is known as flashover.

The rate of heat release is measured by the fire propagation test whilst the surface spread of flame test measures the propensity of a material to spread fire across its exposed surface in the heating conductors of the developing fire. As can be deduced from the description of a developing fire above, these two 'reaction to fire' properties are only two factors in the probability of a small fire developing into a large one.

In the fire propagation test, the surface spread of flame test and the non-combustibility test, the samples are subjected to intense radiation from a radiant source. This is entirely in harmony with developing fire conditions where radiation and temperatures are very high. The fire performance of materials at these high temperatures is far different from their performance at room temperature: they are much more likely to ignite. Therefore, if a fire test is to be really informative, it is essential that it mirrors all these conditions. This is why the 'spectacular' tests of blow lamps playing on materials without them igniting are frowned upon by fire professionals. It also explains why the oxygen index test is falling out of favour. Neither of these tests subjects the sample to very high temperatures.

10.5 The future

Regulations are constantly changing: the UK Building Regulations are constantly being reviewed. It has been suggested that smoke and toxicity should be incorporated into them but this has not so far occurred.

10.6 The European market

The completion of the European internal market will have dramatic effect on building materials. Although this took place at the end of 1992 the actual completion will not be achieved for years: it will be a steady progression to unity. The objective is to reduce all barriers to trade within the European Community. In addition to removing tariff barriers (import taxes) and customs regulations the aim is to remove the technical barriers to trade. These are principally the result of differing product standards and regulations in different countries.

As a first step a *Construction products directive* has been adopted. This directive is designed to bring building laws and regulations in the different countries closer together. The document sets out seven 'essential requirements' with which products must comply. The essential requirements are similar to those embodied in the UK Building Regulations. One of the essential requirements is safety in case of fire. The document defines the requirement as follows:

The construction works must be designed and built in such a way that in the event of an outbreak of fire in the works:

(a)   occupants can leave the works or be evacuated unharmed
(b)   the safety of rescue teams is ensured
(c)   the fire cannot spread to neighbouring works or parts of them
(d)   the generation and spread of fire and smoke within the works themselves is limited.

Therefore the requirement covers life safety and the spread of fire.

Products will be deemed to comply with the requirements by achieving a 'CE' mark (Fig. 10.8). The CE mark will be obtained by certified compliance testing in accordance with a European Standard (EN) or by achieving a European Technical Approval. There are a few European Standards now out; many more will be needed within the next few years. The Comité Européen Normalisation (CEN) is the organisation

European
Community

Communauté
Européenne

10.8 CE mark.

mandated to produce these standards. The CE mark will allow materials to pass from one country to another and be offered for sale. It is not a quality mark, merely a sign of compliance with the essential requirements of a directive.

One of the reasons that the internal market cannot be completed on time is the problem with 'reaction of fire' tests. These differ widely in the various countries of the European Community and the European Free Trade Association and are of course the foundation bricks on which the various National Building Regulations are based. For these reasons no harmonised tests have been arrived at. It appears that the most promising test will be a new test called the 'cone calorimeter'. This test has the merit that it is not presently a standard in any country although it is an international standard test in ISO/DIS 5660.

10.7 The cone calorimeter

Large fire tests are expensive and computer models are beginning to be be devised which, when given certain basic fire performance characteristics of the materials involved and the configurations of the structure, will predict the fire performance of the structure as a whole. Although not yet accepted by regulations, this is the way forward.

Some of the properties which will be the building blocks of this so-called engineering approach are rate of heat release, rate of burning at different temperatures and with different amounts of oxygen, rate of smoke generation and rate generation of toxic gases. The apparatus which looks a contender to supply this information is the cone calorimeter which burns a material when subjected to various predetermined levels of radiation while measuring oxygen depletion, rate of mass reduction, smoke production and toxic gas generation. Much of the development work has been carried out in Scandinavian research laboratories and the Scandinavians believe that the cone calorimeter and a surface spread of flame test will be sufficient. Even after the test procedure has been developed it will take some time before national regulations can be changed. The goal appears to be unified European Building Regulations.

10.8 Designing the Heathrow Terminal 2 walkway panels: Class 0

There were few resins sold as Class 0 resins at the time of the design of the walkways in the early 1970s. It was appreciated that a so-called Class 1 resin was designated to be so after tests on laminate specimens. The glass/resin ratio of the tested laminate had affected the result. The higher the proportion of glass – which is completely non-combustible – the better will be the fire performance since the resin is the combustible ingredient of the composite.

Scott Bader, the resin manufacturers, co-operated with the GRP Fabricators Anmac Ltd to produce one of the first Class 1 polyester resins Crystic 346 PA. For the internal surfaces of the walkway the gel coat was also a Scott Bader polyester resin, Crystic 47 PA. In the surface spread of flame test the gel coat is the prime material under test and so must have good fire properties. However, the modification of a resin to improve its fire properties results in a deterioration in the weathering characteristics and generally precludes its use for outside applications. ANMAC therefore experimented with a number of other coatings for the external surfaces developed their ANMAC An-o-clad system. It is understood that a highly filled fire-retarded gel coat resin was used with a polyurethane lacquer applied over the surface. The lacquer protected the fire retardant gel coat from UV light. Without the lacquer the gel coat would have weathered very quickly and yellowed. This system was very successful achieving a Fire Propagation Index of 6.9 when the maximum allowable is 12 and no flame spread at all in the Surface Spread of Flame test when 165 mm maximum is allowable. Both these tests were carried out on samples with the same make up and resin/glass ratio as the walkway panels. Since the mid-1970s there have been developed a number of different ways to improve resin fire performance: for instance, in the addition of borax, which froths up during the fire, forming an intumescent coating. Alternatively, halogenated waxes are added, which give off chlorine when heated which inhibits flaming. These are almost always used with a synergist, for instance antimony oxide. Antimony trihydrate is often used as a heat-absorbing additive because this undergoes an endothermic reaction in a fire, cooling the flame. Other ways of flame-retarding the polyester resins are adding mineral fillers or modifying the chemistry, forming a HET-acid resin.

10.9 Fire resistance

Achieving a half-hour fire resistance was difficult especially as this was to be the first time that this had been commercially done. First of all the panels had to retain their structural integrity after 30 minutes under test. The design philosophy was that the panels should be double skinned and that the outer skin could be sacrificed in a fire while the inner skin maintained the required structural integrity. The skins were then fabricated 3 mm thick with returned edges to give stiffening. The overall thickness of the panels was 75 mm.

The second criterion was that after 30 minutes under test the inside face should not reach an average temperature greater than 140 °C or 180 °C at any point. In order to achieve this, an insulating material was put into the voids between the two mouldings: a 45 mm thick mat of mineral wool was used. Another of the problems was that if the outside face was sacrificed there was great difficulty in retaining the mineral wool insulation in place. The solution was to use a woven glass fabric (woven rovings) rather than the cheaper and more normally used chopped strand mat which is made up of randomly orientated glass fibres 20–40 mm long. When the resin burned away, the woven rovings cloth remained in place even though the laminate had lost its stiffness. Hence the insulation was retained *in situ*, the temperature controlled and the inner laminate protected.

In the fire resistance test a sample 3 m square was made up and offered to the test furnace. It was required to last 30 minutes. After 60 minutes the test panel was still standing and showed no sign of collapsing and at this time the test was stopped. The unexposed face of the sample attained a temperature of 180 °C at one point after 48 minutes which was the time to failure so that the panel easily achieved the required half-hour Fire Resistance.

In this test a joint had to be tested so the sample had a joint which was to be used in service down the middle of the sample. The joint ranged in width from 10 mm on the outside to 4 mm on the inside. Asbestos rope was inserted into the joint and covered with a two part polysulphide mastic. In the fire test the mastic burned away but the non-combustible rope maintained the seal.

10.10 Conclusions

The objectives of this project were:

1. The exterior surface was to have good weathering properties. Fifteen years after its installation the walkway still retains its original sparkle and shows no sigh of yellowing, crazing or chalking as the photographs taken in 1992 show. Therefore this objective was achieved.
2. The structure had to be structurally sound. The walkways have survived 15 years of constant use with no evident sign of structure distress. Therefore this objective was achieved.
3. The structure had to achieve a 30 minute fire resistance when tested to BS 476: Part 8. Tests showed that it achieved nearly 50 minutes. Therefore this objective was achieved.

4   The internal surfaces had to achieve a Class 0 classification in accordance with UK Building Regulations. The surface easily achieved the Class 1 surface spread of flame and the required level in the fire propagation test. Therefore this objective was achieved.

With all objectives achieved this pioneering project has been an unqualified success.

10.11 Bibliography

10.11.1 British Standards

[10.1] British Standards Institution, *Guide to fire test methods for building materials and elements of construction*, PD6520, London, British Standards Institution, 1988. (Gives a valuable review of BS test methods with commentary.)

[10.2] British Standards Institution, *Fire tests on building materials and structures*, BS 476, London, British Standards Institution.

[10.3] British Standards Institution, *External fire exposure roof test*, BS 476: Part 3: 1975, London, British Standards Institution. (Designed to measure the ability of a representative section of a roof, rooflight, dome light or similar components to resist penetration by fire, when its external surface is exposed to heat radiation and flame. Also measured the extent of surface ignition.)

[10.4] British Standards Institution, *Non-combustibility test for materials*, BS 476: Part 4: 1970 (1984), London, British Standards Institution. (Includes definition of non-combustibility. Determines whether materials, with or without coatings, used in construction or finishing of buildings meet the definition.)

[10.5] British Standards Institution, *Method of test for ignitability*, BS 476: Part 5: 1979, London, British Standards Institution. (Measurement of ignitability characteristics of essentially flat and rigid building materials and composites when subjected to a small flame.)

[10.6] British Standards Institution, *Method of test for fire propagation for products*, BS 476: Part 6: 1981, London, British Standards Institution. (A method to determine the fire performance of products used as internal linings in buildings. Takes account of the combined effect of factors such as the ignition characteristics under intense heat radiation, the amount and rate of heat release and thermal properties of products in relation to their ability to accelerate the rate of fire growth.)

[10.7] British Standards Institution, *Method for classification of the surface spread of flame of products*, BS 476: Part 7: 1987, London, British Standards Institution. (Method for measuring the lateral spread of flame along the surface of a specimen in the vertical position under intense heat radiation together with a classification system based on rate and extent of flame spread.)

[10.8] British Standards Institution, *Guide to the principles and application of fire testing*, BS 476: Part 10: 1983, London, British Standards Institution. (Describes the general principles and application of methods in this series of standards for fire testing of building products, components and elements of construction.)

[10.9] British Standards Institution, *Method for assessing the heat emission from building materials*, BS 476: Part 11: 1982, London, British Standards Institution. (Describes a method for assessing the heat emission from building materials when inserted into a furnace at a temperature of 750 °C. This test is based on the same principles as BS 476: Part 4.)

[10.10] British Standards Institution, *Method for determination of fire resistance of elements of construction (general principles)*, BS 476: Part 20: 1987, London, British Standards Institution. Also ISO 834. (General details of test conditions, specimens, apparatus and criteria for fire resistance testing. Appendices give general guidance information, method for residual loadbearing capacity and operating instructions.)

[10.11] British Standards Institution, *Methods for determination of the fire resistance of loadbearing elements of construction*, BS 476: Part 21: 1987, London, British Standards Institution, 1987. (Detailed test requirements for beams, columns, floors and flat roofs and walls.)

[10.12] British Standards Institution, *Methods for determination of the fire resistance of non-loadbearing elements of construction*, BS 476: Part 22: 1987, London, British Standards Institution. (Detailed test requirements for partitions, fully insulated, partially insulated and uninsulated doorsets and shutter assemblies, ceiling membranes and glazed elements.)

10.11.2 UK British Regulations 1991

[10.13] *Mandatory rules for means of escape*, London, HMSO. (Applied to buildings (including dwelling houses and flats) of more than three storeys.)

[10.14] *Approved Document B, Fire Safety*, London, HMSO. (Gives suggested solutions to fire problems in buildings.)

10.11.3 Cone calorimeter

[10.15] Babrauskas, 'Development of the cone calorimeter – a bench scale heat release rate apparatus based on oxygen consumption', *Fire and Materials*, 1984 **8** (2) 81.

[10.16] Huggett, 'Estimation of rate and of heat release by means of oxygen consumption measurements', *Fire and Materials*, 1980, **4** (2) 61.

[10.17] Parker, J., 'Calculation of the heat release rates by oxygen consumption for various applications', *Fire and Sciences* 1984, **2** 380.

10.11.4 Other publications

[10.18] L. Hollaway, (ed.) *The use of plastics for loadbearing and infill panels*, London, Manning Rapley, 1976. (Chapter 9 describes a fire test on building panels to simulate extreme fire conditions.)

[10.19] *Modern Plastics Encyclopedia*, pp. 658–663, New York, McGraw-Hill, 1986–7. (Gives a comprehensive list of flame retardants for use in plastics.)

[10.20] A.J. Legatt, *GRP and buildings*, London, Butterworths, 1984. (Chapter 7 gives suggestions for 'real fire' situation testing, a short section on fire tests and hints on fire design.)

[10.21] H.V. Boeing, *Unsaturated Polyesters: Structure and properties*, London and New York, 1964. (Chapter 8 gives an explanation of the various means for introducing flame retardancy to thermoplastics.)

11 Design of a commercial vehicle sideguard system

11.1 Introduction

In 1984 the United Kingdom Department of Transport introduced a new legal requirement to fit side underrun devices to newly registered commercial vehicles (Fig. 11.1). Tests had shown that beams fitted to the sides of vehicles prevented cyclists, motorcyclists and small cars being run over by the rear wheels during an accident.

The government requirements contained detailed and rigorous geometrical, strength and stiffness parameters which had to be met by any system. The outside face of the beam arrangement had to be smooth, the beams had to fit the space between floor, road and front and back wheels leaving not more than specified maximum gaps. The system had to withstand design forces of 2.0 kN applied anywhere, and should not deflect more than specified values.

Vehicle operators approached Maunsell to see if it was possible to produce a lightweight composite system that would minimise the extra weight that vehicles would have to carry. Extra weight means increased fuel costs and reduced payloads. A rapid feasibility study showed that a pultruded glass reinforced plastic beam system would be likely to satisfy the requirements of the legislation and would also be light enough to be attractive to clients. It was also concluded that the cost should be attractive to clients seeking weight-saving solutions. A group consisting of the designer Maunsell Structural Plastics, a pultruder and a vehicle body manufacturer agreed to work together to bring a system to the market-place within six months. It was to be called SIDESAFE.

11.2 Design development

There was neither the time nor the cash available to build a prototype structure, which had been normal practice hitherto for structural applications of composites. On the other hand the system had to meet the rigorous legal performance requirements. Also the fixings had to be configured to allow ease of installation on a whole range of vehicle

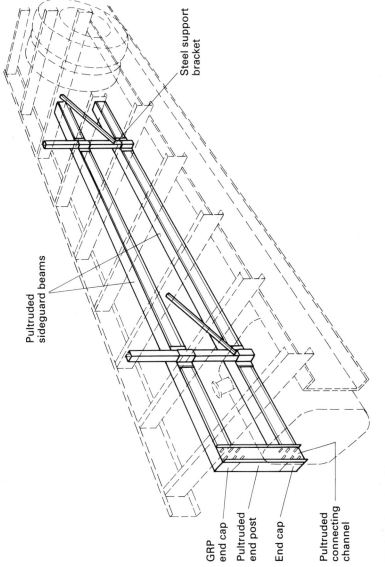

Steel support bracket

Pultruded sideguard beams

GRP end cap

Pultruded end post

End cap

Pultruded connecting channel

11.1 Pultruded GRP sideguard device.

sizes and types. Maunsell had been developing a limit state design method for reinforced plastic materials and it was considered to be sufficiently well established to be used for the first time to design the SIDESAFE system.

A critical first decision was to decide upon the number of supports that would be used to attach the beam system to the large range of different rigid chassis and floor structures. It was decided early on to use only two basic supports in order to minimise fixing time. A pultruded glass reinforced plastic box beam component 100 mm deep was quickly shown to be adequate to resist the required loads with acceptable deflections when spanning between supports and cantilevering at the ends. However, the key to a successful design was to meet the following objectives:

1 Achieve a minimum material content and lowest possible pultrusion cost.
2 Have simple fixing details requiring no drilling, welding or bonding of components – also minimise the fixing costs.
3 Achieve good durability, particularly with respect to vibration and fatigue at fixing positions. The system would be subject to the most rigorous vibration, abrasion and other environmental conditions.

It was realised that these objectives would not be achieved unless a very sophisticated approach was taken to overall analysis and detailed design.

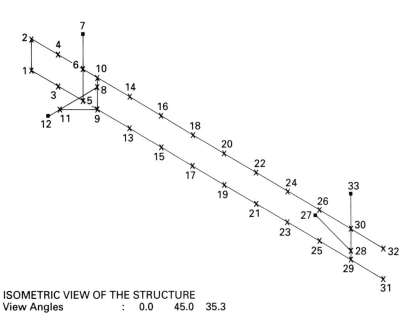

ISOMETRIC VIEW OF THE STRUCTURE
View Angles   :   0.0   45.0   35.3

11.2 Finite element computer model of structure.

11.3 Structural analysis

The beam structure was modelled using a three-dimensional space frame structure using a finite element computer program (Fig. 11.2). The model contained both GRP pultrusions and steel support structures, with the chassis and support floor assumed to be rigid. Design loads were applied to the model in all directions as required by the legislation, and maximum forces in all the connections were evaluated under both static and dynamic loading. Maximum deflections at all locations were checked against limiting values. Pultruded beam stiffnesses were calculated by the use of an in-house computer program, LAMINA, which calculates the laminate properties for each part of the beam, taking account of the proposed glass fibre configurations.

It was clear from the first analyses that a GRP system would be more flexible than a steel system and therefore great care had to be taken in the detailing of connections to the rigid structure to avoid local stress concentrations and the possibility of progressive fatigue failure in the

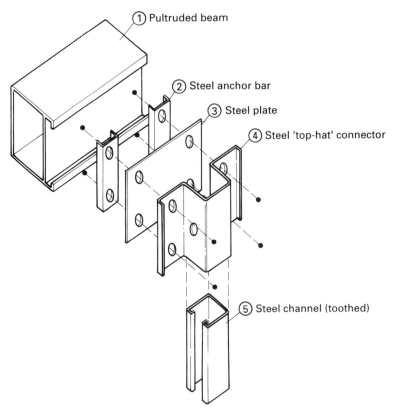

11.3 Connection detail.

laminates adjacent to the connections. This problem was overcome by the design of a unique system clamping onto an open steel channel support (Fig. 11.3). This channel, having low torsional stiffness, would allow the GRP beam system to rotate at support points without generating high local stresses. This behaviour was evaluated using the finite element analysis. Both longitudinal and lateral forces were applied to the model in order to produce an envelope of maximum forces on all the components and connections.

11.4 Design of pultruded beams using the limit state design method

A considerable amount of optimisation was carried out to derive a pultruded box section shape and glass configuration which would achieve the design objectives with minimum material content. The beam that was eventually selected (Fig. 11.4) had different wall thicknesses on each side and lugs to permit the clamping attachment at any location along the length. The box section was designed to have a high glass content and to use an isophthalic polyester resin. Longitudinal rovings provide longitudinal stiffness and continuous filament mats in each laminate face provide the transverse strength and integrity. The following sections give the primary design aspects.

11.4.1 Beam properties

Moment of inertia I_{yy} = 884 500 mm^4
Section moduli z = 26 364 mm^3 (centre of front wall)
19 787 mm^3 (extreme fibre at rear)
Transverse shear area A_x = 300 mm^2

The laminate forming the front wall is three-ply. The predicted properties of this laminate (calculated entirely from laminate theory) are as follows:

Plate (flexural) rigidity matrix:

D_{11} = 77 876 N mm (longitudinal)
D_{22} = 40 612 N mm (transverse)
D_{12} = 12 161 N mm
D_{33} = 15 967 N mm (torsional)

Equivalent longitudinal in-plane modulus:

E_{11} = 29 539 MN/m^2

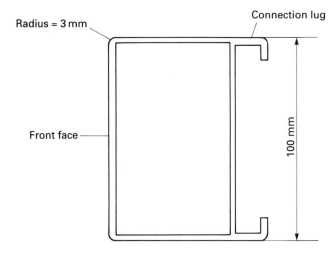

11.4 Pultruded GRP beam cross-section.

For the laminate forming the shorter walls, the predicted equivalent in-plane shear modulus is:

$$G = 4500 \text{ MN/m}^2$$

The above values were adopted as characteristic material properties for preliminary design.

11.4.2 Limit states

The critical limits are judged to be:

1   Deflection at tip of rear cantilever.
2   Local buckling (wrinkling) of front wall at midspan.
3   First damage at extreme rear fibre at midspan.

Since the front wall is expected to exhibit substantial post-buckling strength, the above limit states are all serviceability criteria.

11.4.3 Partial coefficients

The design loading is a one-off test loading applied under controlled conditions, therefore the partial load coefficient, γ_f, is taken as 1.0.

The partial material coefficient allows for uncertainties arising from the production process and in the method of assessing the characteristic values. Large factors are not required at serviceability limit states, and the pultrusion process provides a highly consistent and accurate product. However, the characteristic properties are derived from theory rather than test and there is therefore uncertainty associated with this, particularly for shear rigidity values where experience has shown the predicted values to be optimistic. For SLS γ_m is therefore taken as:

$$\gamma_m = 1.1 \text{ (all values except shear, torsion)}$$
$$= 2.0 \text{ (shear, torsion values)}$$

11.4.4 Material property reduction factors

There is no sustained or repetitive loading and the environment is normal. Only age degradation is applicable, but the pultruded beam incorporates a surface veil for durability and experience has shown little degradation of stiffness with time. No property reduction factors are therefore applied.

11.4.5 Design calculations

11.4.5.1 Deflection at tip of cantilever
Allowing for shear deformation, deflection (Fig. 11.5):

$$d = PC^2 (C + L)/3EI + PC/GA$$

where L = span of beam, C = cantilever overhang, P = load at end of cantilever:

$$
\begin{aligned}
EI &= (E_{11}/\gamma_m)I_{yy} \\
&= (29\ 539/1.1)\ 884\ 500 \\
&= 2.375 \times 10^{10} \text{ N mm}^2 \\
GA &= (G/\gamma_m)A_x \\
&= (4500/2.0)\ 300 \\
&= 675\ 000 \text{ N}
\end{aligned}
$$

therefore, deflection:

$$d = \frac{2000 \times 500^2\ (3000+500)}{3 \times 2.375\ 10^{10}} + \frac{2000 \times 500}{675\ 000}$$
$$= 26.00 \text{ mm}$$

that is, less than the permitted 30 mm.

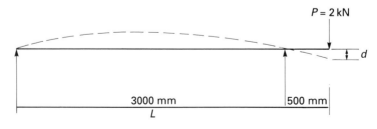

11.5 Deflection at tip of cantilever.

11.4.5.2 Local buckling of front wall at midspan

In the view of the expected substantial post-buckling strength, and the limited number of load cycles expected, the stress will be permitted to exceed the onset of buckling stress by 25%. Within this range the buckling deformations are expected to be elastic and fully recoverable on removal of load. Subsequent testing showed the post-buckling strength to be 120 MN/m². Assuming the plate is simply supported at the edges, the critical buckling stress f_{cr} is given by:

$$f_{cr} = (2\pi^2/b^2t)(\sqrt{D_{11}D_{22}}) + (D_{12} + 2D_{33})$$

b is plate width (97.5 mm), t is plate thickness (3.5 mm) and D are plate flexure rigidities. Therefore applying γ_m factors:

$$
\begin{aligned}
f_{cr} &= (2\pi^2/97.5^2 \times 3.5)\,(\sqrt{(77\ 876 \times 40\ 612/1.1 \times 1.1)} + \\
&\quad 12\ 161/1.1 + 2 \times 15\ 967/2.0) \\
&= 46.4\ \text{MN/m}^2
\end{aligned}
$$

Bending moment on cross-section at midspan:

$$
\begin{aligned}
M &= PL/4 \\
&= 1.5 \times 10^6\ \text{N mm}
\end{aligned}
$$

Therefore compressive stress at centre of plate:

$$
\begin{aligned}
f &= M/z \\
&= 1.5 \times 10^6/26\ 364 \\
&= 56.9\ \text{MN/m}^2
\end{aligned}
$$

which exceeds f_{cr} by 22.6%.

11.4.5.3 First damage at extreme rear fibre at midspan

The first damage criterion will be taken as a maximum strain limit of 0.3%. Bending moment at midspan is $1.5*10^6$ N mm, therefore tensile stress at extreme rear fibre:

$$
\begin{aligned}
f &= M/z \\
&= 1.5* 106/19\ 787 \\
&= 75.8\ \text{MN/m}^2
\end{aligned}
$$

The strain is given by:

$$
\begin{aligned}
e &= f/(E_{11}/\gamma_m) \\
&= 75.8/(29\ 539/1.1) \\
&= 0.0028
\end{aligned}
$$

which is less than the permitted 0.003.

11.5 Manufacturing specifications, tooling and production

In parallel with design, a detailed manufacturing specification was prepared, in conjunction with the pultruder, so that the material variations assumed in design would be matched in the product. For example, wall thickness variations were limited to ±0.25 mm. As soon as the pultrusion geometry and glass configurations were finalised, tooling manufacture and glass infeed system design were commenced in parallel. Tooling manufacture took five weeks and within five months of the start of the project the first pultrusions were produced to the exacting specification. These were some of the first pultrusions to be produced to a detailed structural engineering specification. The first beams were then assembled into a complete system and taken to the Motor Industries' Research Association (MIRA) where they were tested to ensure system compliance with the legislation.

11.6 Proof testing of the system

MIRA were testing a number of systems in a special rig and SIDESAFE was fitted into a slot in the test programme. This was the first time the system had been attached to a vehicle.

Test requirements were that the sideguard must be capable of withstanding a force of 2 kN applied perpendicularly to any part of its surface by the centre of a circular loading face not more than 220 mm in diameter (Fig. 11.6). During this application, deflection must not exceed 150 mm or 30 mm at any point less than 250 mm from the rear

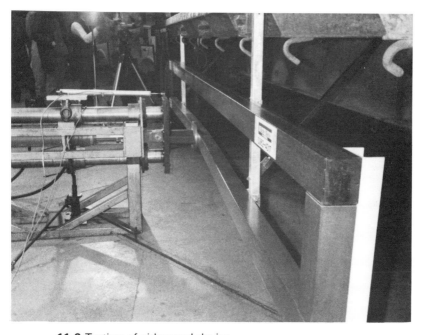

11.6 Testing of sideguard device.

edge of the guard. Loads were applied in six locations and in two directions, laterally and longitudinally. In a final test, loads were increased to induce local buckling and then further to induce complete failure of the system to produce the ultimate capacity.

Extremely good agreement was found between the test results and the predicted behaviour from the finite element model, as follows:

Equivalent in-plane modulus (predicted values in brackets):

$$\text{Test } E_{11} = 28\ 010\ \text{MN/m}^2\ (29\ 539\ \text{MN/m}^2)$$
$$\text{Critical buckling stress } f_{cr} = 56\ \text{MN/m}^2\ (59.5\ \text{MN/m}^2)$$

Thus the system met the requirements of the legislation and was certified for use.

11.7 In-service performance

System units were being sold within the six months' target. Several thousand sets of sideguards have been installed on commercial vehicles and have successfully undergone many thousands of vehicle miles. The condition of beams inspected after five years' continual use on articulated trucks was excellent. No problems were encountered at the fixings and the surface condition of the pultrusions was almost as new. The polyester surface veil clearly gave excellent durability. The only reported problems were in beams which had hit traffic bollards during parking. They were visually fine, but on inspection the impacts had split the box sections at the corners. This highlights a key problem with composites: damage detection if the sections are abused in accidents. This problem is particularly acute on hired vehicles where insurance may not cover damage unless it is reported at the time.

The design, manufacture and performance of the system met all expectations. Indeed, it was concluded that the performance was more predictable than expected from steel and aluminium sections because of more accurate fabrication and the lack of locked-in temperature stresses. However, the design was a complex and demanding process, particularly to achieve a cost-effective end product, and needed the highest level of structural engineering skills.

12 Design and production of a composite chassis component in the automotive industry

12.1 Introduction

Despite the demonstration of safety-critical structural applications of polymers in the automotive industry, e.g. the leaf spring, the lack of commercial and technological experience in using glass fibre reinforced polymers (GFRP) restrict their growth. In an attempt to close this knowledge gap, Rover has established a multidisciplinary composite team developing components in collaboration with Rover engineers. The design experience, engineering disciplines and data requirements for producing a large GFRP structural component is described and the chapter illustrates the development procedures required by the industry.

The increasing lobby for the car to become acceptably 'green' has led to a renewed vitality in the investigation of polymers and composites as lightweight replacements for heavy metal items. Although there is a wide usage of polymers in the automotive industry, applications are restricted to non-safety-critical components; the majority of use is in the thermoplastic injection moulded trim items. Increasing the use of polymers in structural components is promoted by many material suppliers, especially in the areas of fibre reinforced polymers.[12.1] Prototype component steering wheels,[12.2] steering columns[12.3] and seats[12.4] are frequently being demonstrated, but the cautious application of these developments is a constant source of frustration to the suppliers, who consider the automotive's long term investment in steel as the biggest inhibitor to polymer growth.

The metal domination is particularly relevant when considering change with the automotive engineer, who is not only steeped in traditional metal design, but knows that the infrastructure, such as supply links, tooling, design analysis, test and cost programmes will be supportive and the characteristics of metal will be available in information data banks; this is not necessarily so for polymers. Consequently change is slow. To establish use of the composites in the mass auto market does not just rely on a conviction that this material is beneficial in design, performance, quality and cost, it also requires a full movement in the technology disciplines that contribute to the manufacture of a car – probably involving more technologies than any other industrial product.

In order to create a material design culture that includes composites alongside metal within Rover, a composites team was established at the Advanced Technology Centre, University of Warwick, to include Rover engineers in its development of real components for the future vehicle, introducing these materials into a metal environment using the most modern technologies. The automotive composites design methodology is maturing from this system.

12.2 Development strategy

In order that Rover maintains quality and reliability in its vehicles, there is a high carry-over into new vehicles of components, systems and designs which have been proven in the field. The concept is one whereby systematic research, high integrity design and exhaustive testing of new products and concepts are carried out in advance of application to a particular model.

Selection of a suitable component for introduction of GFRP in the vehicle in line with this system has resulted in work on a current vehicle at the front end of its production life, Rover 800, for which a composite part replacement can be proven in prototype vehicle trials as well as on a test bed. Thus, a composite component will be considered to evolve from a successful, well-tried metal design, rather than a complete design change.

12.3 Component selection and feasibility

The chosen component had to be one of replacement for a current metal one in order to meet Rover policy of proving the technology before commercial application. Feasibility studies suggested a chassis component as a challenging high stress item enabling a considerable number of design assurances regarding the technology to be assessed. The selection for the purposes of application to a vehicle required a car component at the front end of its development 'life' and a component well-proven in service – a Rover 800 rear beam of the front suspension (Fig. 12.1) – was chosen.

This component, like many in the automotive industry, is designed with stiffness as a major factor, and GFRP materials have lower flexural modulus than steel. However, the following material comparisons illustrate that although this results in thicker GFRP sections than steel, composites can still provide weight savings (Fig. 12.2). Thicker sections mean costs can rise, so it is recognised that successful introduction of GFRP

12.1 Rover 800 rear beam (front suspension).

material will only be achieved with the correct design that takes advantage of GFRP's ability to integrate a number of metal parts into one functional component. By such, integration costs can be controlled within the requirements of a high volume industry.

The major determining factor in the selection of a structural metal chassis component (Fig. 12.3) for development in composites was the number of steel pressings in the assembly that could be integrated into a one piece moulding (12 : 1 ratio). This part agrees with a belief in the automotive industry that over ten parts must be integrated into one for commercial success. Parts integration enables savings on tools, jigs and assembly and a cost saving of 30% is being considered.

This saving is an attractive incentive but other objectives were of equal importance – establishment of a *commercially viable, structural* component for *high volume manufacture* and capable of meeting the *quality* demanded in a *safety-critical* engineering application.

The high volume GFRP process considered applicable to such a structural application of high stiffness and mechanical loading is resin injection moulding (RIM) of a glass preform. As a low pressure process, RIM also has the potential for lower tool costs.

Interestingly, similar complex shaped preforms for chassis components using RIM are being reported by other car manufacturers.[12.5, 12.6] These reports appear to announce a concept engineering development rather

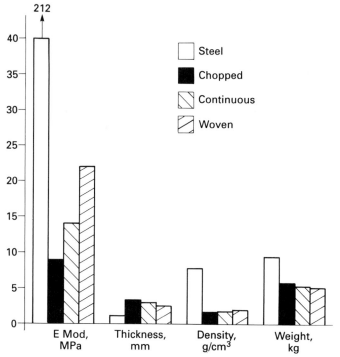

12.2 Comparison of material laminates for equal design stiffness.

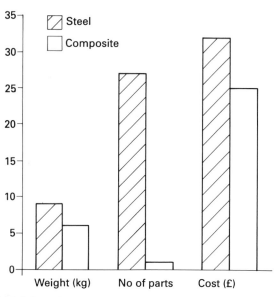

12.3 Benefits of parts integration.

than claiming production capability, they contain stitched or hand laid glass preforms which suggests the development is a long way from commercial viability. The nearest to market application must be one reported by Peugeot[12.1] who use the thermoformed glass preform similar to the one outlined in this chapter.

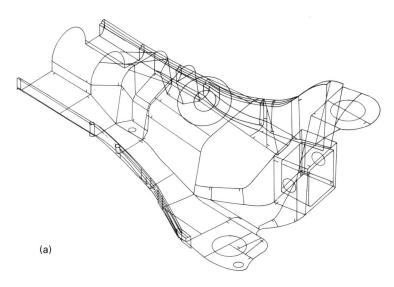

(a)

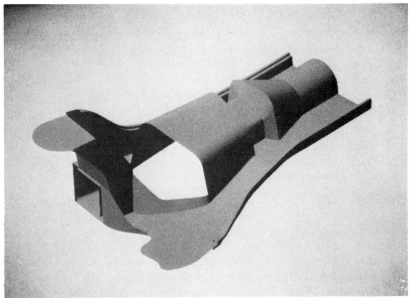

12.4 CAD design of GRP subframe: (a) wire frame CAD and (b) CAD surface model of fibre reinforced composite chassis component.

12.4 Component design

As a result of 'in-house' discussions, together with the overriding situation that the part was a replacement one, the GFRP design could not adopt any revolutionary shape. It still had to fit the current space, meet fixture points and align with other components. Thus the GFRP design proposal maintained a complex shape (Fig. 12.4), not ideal for moulding but a best compromise.

The GFRP rear beam design consists of two parts, an upper and lower section bonded together; it is approximately 1000×400 mm and of complex shape to meet packaging in the car. The initial design is based on epoxy resin impregnated continuous fibre (Vetrotex Unifilo); thermoformed continuous fibre is considered to be foremost in available preform technology. Epoxy resin was selected as a proven system in structural applications, for instance, aerospace structures and, nearer to home, leaf springs.

Computer-aided design (CAD) usage throughout the automotive industry is necessary for data transfer of the part design to all relevant engineering sections. A model is created in 3D wire frame and surfaced. It was essential that this was undertaken with the component engineer familiar with design rules who could advise and test spatial interference problems.

12.5 Design analysis

Simple calculations from flat plate theory indicated a component thickness of 5 mm with a material containing 50% glass by weight would be adequate to satisfy the worst load case. No data existed at this stage that were sufficient to enable the automotive design analysts to model the component accurately. RIM epoxy impregnated test pieces were made to the proposed material specification and data was accumulated for the following properties:

flexural modulus (X, Y, Z);
Poisson ratio (X, Y, Z);
shear modulus;
ultimate tensile strength.

This enabled design analysis to be undertaken for the epoxy impregnated Unifilo glass using the finite element computer software ANSYS. The work initially validated the data and system using a flat plate specimen and then confirmed a 3D component by testing the data on a bumper armature made with a similar GFRP. The full GFRP rear beam component could then be analysed on the system for 12 load cases. The most arduous load is in the area of the lower arm, which, under pot hole breaking, must survive repeated loading to 14 kN. A simple GFRP sub-assembly (Fig. 12.5) was made and

12.5 Sub-assembly of the rear beam lower arm fixture.

assembled for test. Repeated loading of this fixture to 20 kN provided confidence in its design performance.

12.6 Numerically controlled tool production

One of the advantages of RIM is that since it is a low pressure process there may be the opportunity to use low cost tooling for production. For this reason the tooling procedures were of interest to the group, even at prototype stage. A standard manufacturing procedure was used to obtain machining data from the design so as to program a numerically controlled (NC) five axis machining centre to cut the tool. Once programmed, this facility provided pattern, jigs or tools repetitively; even so, with time pressing, the simplest approach appears to have been the five axis machining of a Ureol tool, which could be used as masters to provide GFRP patterns (incorporating relevant shrinkage criteria) to subsequently make kirksite tools. The gating requirement was advised by resin collaborators, Shell, who based their information on a computer flow simulation of the exposy resin path through glass preform. After using the Ureol master to produce the metal tool it became a tool for preform manufacture. It had to be modified by skimming 5 mm thickness from the peripheral surface in order that glass could be press preformed, and yet there was space for the excess glass to overhang as illustrated in Fig. 12.6.

The Ureol tool was capable of cold-press moulding in order to provide initial component samples but metal was essential for an authentic prototype tool especially since it was intended to undertake RIM at 120 °C.

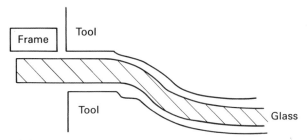

12.6 Schematic representation of the periphery of the Ureol master.

12.7 RIM manufacture

12.7.1 Preform

The Ureol tool was mounted in the press and, although the press was somewhat slower than recommended, adequate preforms were made according to the following schedule:

1 Heat preform 2 min.
2 Transfer time 10 s.
3 Press time 30 s.
4 Preform removal.

It is to be noted that, although this worked at this stage of the development, faster times are achievable with dedicated manufacture. At the time of reporting, five layers of glass have been press-formed in the Ureol tool and these quantities have been doubled in the metal tool when additional pressure could be applied.

12.7.2 Trimming and preform

The trimming of the preform was found to be a critical aspect of the RIM process in order to have accurate placement and fitting of the preform in the tool. A simple jig was tried by thermoforming polycarbonate sheet in the Ureol tool but, despite careful cutting, further individual fitting in the metal tool was needed to achieve adequate accuracy of fit. Thus, a further jig is required for cutting purposes and enhances the value of using an NC program for the first tool generation.

A worrying commercial aspect is the amount of waste glass generated in the process (Fig. 12.7); for the top section, the weight of wastage in this production was at least equal to the component preform and it was difficult to envisage any reduction. Processors[12.7] claim that this can be overcome by usage of the waste offcuts for components requiring smaller glass profiles. This means that the process demands a special-

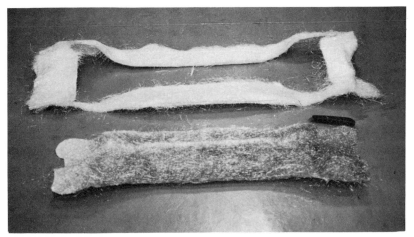

12.7 Preform and trimmings.

ist company with a wide spectrum of different size components in manufacture that can be matched for size.

More interesting to the automotive industry will be the avoidance of waste, hence spray glass preform manufacture[12.8] has been suggested as one example of a technique to reduce waste.

12.8 RIM

If a well pressed and neat fitting preform is available, RIM manufacture is a well controlled process. Nevertheless, to ensure that the resin fill proceeds gradually from injection port to the back of the mould a method of expelling air from the advancing resin front is required. Our collaborator, Shell, hold a patented process for resin injection into preform using a special mould-sealing method to ensure the elimination of small voids, in particular, entrapped air. This RIM process requires that the press is closed with a foam strip at the periphery of the tool, which allows air expulsion but seals when resin contacts it. This produced a satisfactory and void-free moulding. The advice on gating in the tool was demonstrated to be remarkably accurate for flow path prediction in these first off mouldings (Fig. 12.8).

The selected epoxy resin and hardener system undergoes rapid cure after mixing and, hence, requires to be mixed and injected in the tool simultaneously. This speed of injection is the essence of resin injection moulding (RIM) rather than resin transfer moulding (RTM) where the mixed resin and hardener has longer pot life and is transferred slowly. At 120 °C, the epoxy RIM system cures rapidly and enables the finished component to be demoulded in 3 minutes. This cycle time is required by the project plan to achieve the high volumes necessary.

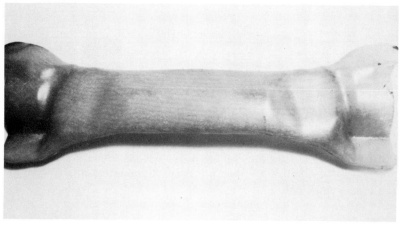

12.8 Flow paths in GFRP rear beam top moulding.

These times have been achieved at the Rover Advanced Technical Centre even though a much slower injection time (20 s) is used than is demonstrated with the RIM machine at Shell (2 s). The implication is that even faster cycle times could be optimised. A comparison of the two techniques is still to be concluded.

12.9 Process conclusions

The moulding of a RIM component within Rover has helped to demonstrate the high volume applicability to structural items as well as providing all the ancillary detail required to make the process a designer's tool at concept. Hence, as far as RIM is concerned, the major conclusions are:

1 RIM is capable of providing the structural requirements in a cost efficient, volume manufacturing process. The data have been obtained in this project for such a design exercise.

2 Tooling must include separate preform tools and injection tools and, moreover, there must be provision for a preform cutting system as well as jigs for any post-machining. However, none of these needs to be expensive.

3 Preforms can be made to complex shapes; however, the efficiency of the process is dependent on the preform cutting profile and the accuracy of the cutting process.

4 Waste offcuts are a major consideration.

12.10 Prototype and further work

This component has been engineered and produced in a commercial and high volume design situation such that a product is now ready for vehicle test.

Since each of these composite parts will be a part of a larger assembly to which the part has to fit or be attached in some way, this means it will probably require machining to a tolerance with various assembly fixings (bolts, adhesives, inserts, etc.) to be added. Each fixing has to be designed for the load and tested to an agreed specification. Methods of design analysis need to be applied and validated. A failure mode and effect analysis (FMEA) will be undertaken to ensure that the subframe manufacture, assembly and application will meet the required quality and safety standards. In addition a non-destructive test programme is required to monitor its continued acceptability during high volume production. These concerns have led to suggested further project areas for investigation to provide the necessary background to confident composite designs to match designing in metal:

machining
tolerance
fixing
adhesive
test
design data library
valuation
assembly

The composite subframe may not be applied in the near future to a market vehicle, but it is enabling the engineers to acquire the design methodologies with this different technology so they can feel confident with their application in the future.

12.11 Design composite storage and access using knowledge based systems

The data that have been accumulated in the component project cover details of design as well as many factors in composite manufacture, e.g. cost and process times, hence, there is a need for efficient data storage systems detailing applications of a composite manufactured component. Such storage databases are becoming available through intelligent knowledge based systems (IKBS). Information gained from this exercise and, indeed, recent attempts to improve the data available to engineers[12.9] will only become effective if, when added to the vast amount already on the engineer's shelf, it can be computer accessed and manipulated using informed expert knowledge intelligently. Artificial intelligence systems are being developed for this application with ATC in order to provide a composite design methodology to the concept engineer of the future.

12.12 References

[12.1] A. A. Adams, A. Bennett, G. F. Heron, G. F. Smith and A. J. Wootton, *Overseas Science and Technology Expert Mission (OSTEM) Fibre Reinforced Plastics*, NEL Publication for DTI, 1987, *Vetrotex Fiberworld* June 1989, 18.

[12.2] 'Plastic springs on the road', *British Plastics and Rubber* January 1986, 4.

[12.3] 'German auto centre looks at composites', *Plastics and Rubber Weekly*, February 8 1986, 9.

[12.4] *Glass Fibre Composites in European Automobiles and Trucks*, Owens Corning Publication, p. 17, 1987.

[12.5] C. F. Johnson, N. G. Chauka, R. A. Jeryan, C. J. Morris and D. A. Babbington, 'Design and fabrication of a HSRTM crossmember module', in *Proceedings of the Third Advanced Composites Conference* (Detroit, MI), ASM International, September, p. 197, 1987.

[12.6] R. D. Farris, 'Composite front crossmember for the Chrysler T11 mini van, in *Proceedings of the Third Advanced Composites Conference* (Detroit, MI) ASM International, September, p. 63, 1987.

[12.7] The Satge, Cofim, Chambery, Frane, Private Communication.

[12.8] 'Fibre directed for a better preform in RTM Usage', *Plastics and Rubber Weekly*, 4 March 1989.

[12.9] F. J. Lockett, 'The way forward', BPF Design Data Conference, October 1988.

13 The use of pultruded profiles in a structural application

13.1 Introduction

Pultrusion is a manufaturing process used exclusively in the reinforced plastics industry. It came into existence in the early 1950s, and was originally very crude. It exhibited very good potential as a mass production technique but for many years that potential could not be attained. The process appeared to be simple and technically not very demanding. But in spite of this no great strides were achieved until quite recently when various manufacturers began to look critically at the process and obtained support from academic institutions in this work. The complexity of the process is now appreciated and advances made towards understanding it. The work to date has changed pultruded profile out of all recognition. It is now highly controlled and of consistent quality and exhibits remarkable properties which justify its use in many diverse and demanding applications.

The designer of composite systems has available an extraordinary degree of variety in choice of fibre, fibre geometry, matrix and amount of fibre present. This results in an enormous range of potential properties. In order to achieve the most effective design the designer has the option to optimise not only the construction but also the shape of the component, by for instance, putting the fibres in the direction of the principal stresses. Thus every design can have a different construction depending on the stresses being applied. For mundane applications, the amount of work required would be prohibitive. This problem is alleviated by choosing between the two available design approaches. It can be a selective process by the use of prescriptive design codes or it can be an analytical process and thus allow an optimum solution to be determined.

In the selective design process the design brief stipulates or implies a particular area of search. Once a solution has been found very little or no attempt is made to optimise it. This is the case when designing with standard structural elements. 'I' beams for example. Particular sizes are available from which to select the most appropriate element. In the selective design process there is no opportunity to refine or optimise, but it provides rapid economic solutions to common engineering problems.

Composite structural elements are available as standard pultrusions in the form of box beams, angles, channels, etc. The properties and performance are known, thus allowing structural analysis to be carried out. Hence designs may be made on a selective basis from these 'off-the-shelf' composites. On the other hand, composites in general and pultrusions in particular may be 'tailor-made' for a particular design brief thus allowing an optimised design to be achieved. In the case of optimum design of composites not only is the geometry (shape) designed but also the material itself. Material design and geometry cannot be solved in isolation. Candidate solutions are analysed and compared. The process is iterated until the optimum solution is found.

The use of classical laminate analysis of composite materials is discussed elsewhere.[13.1] This is a powerful tool for the analysis of composite materials including those that are pultruded. It allows materials to be modelled consisting of multi-layers of differing fibre reinforcements at any angle. The elastic properties may be predicted, and with the application of a failure criterion, the failure can be predicted.

Performance criteria determine whether a standard structural element will be a suitable solution for a particular design brief. If not, then a custom shape must be designed. The geometrical shape of a pultrudate is a function of the performance required in service and the restrictions imposed by the production process. The critical aspects of pultrudate geometry are thickness, radii, overall size, number of cavities and location of the reinforcement layers.

Glass fibre pultrusions have a significantly lower tensile modulus than steel or aluminium. This results in the design procedure being concerned initially with the deflection limits and only subsequently with the strength requirements which are generally found to be satisfactory. The relatively low modulus of glass fibre pultrusions also requires that buckling characteristics be assessed a little more critically than would be the case with a steel component which have an abundance of stiffness relative to strength. The shear stiffness of composite materials is relatively low. This can give rise to deflections in beams due to shear which are appreciable, and can be of similar magnitude to those due to bending. It is therefore essential to determine by calculation the values of these deflections to ensure they are insignificant or that they are taken into account. This is particularly important with deep sections, short spans and hollow sections.

Composites do not exhibit a yield zone in their stress–strain response. Therefore 'local yielding' cannot be relied upon to solve certain design problems, as is the case with steel design. Stress concentrations, perhaps because of misalignment of elements, cannot be left to yield as might be the practice with steelwork. The stress will remain in place and may result in premature failure. A further result of the lack of a yield zone in composites is a requirement for the use, for instance, of large washers with bolted connections to allow local stresses to be spread more evenly.

13.2 An example of selective design

In order to illustrate the solution to a typical example of the use of pultruded elements in a structural engineering application, a simple example has been selected which is complete in itself (see Fig. 13.1). The selected structure is an indoor platform to allow access over a chemical process tank. The tank contains sulphuric acid and it can be assumed that this will inevitably result in some of the acid gaining access to the surrounding process area. Hence the environment is aggressive.

The platform is about 3 m long by 1 m wide and is supported on four columns each 3 m high. The flooring medium will consist of a pultruded grating system which is required to carry a live load of 4500 N/m². At this load the grating deflection must not exceed 6 mm. The structural elements must not deflect by more than span/180. A hand rail is required on two parallel sides of the platform and as it is required for occasional use only, a vertical ladder is suitable. All joints are to be bolted and epoxy bonded.

13.2.1 The grating

The fire-retardant grade polymers used in the gratings of a typical supplier (e.g. [13.2]), are suitable for the chemical environment of up to 10% sulphuric acid concentration at 65 °C. If it is assumed that the longitudinal structural elements will be, say, 150 mm wide and that the grating spans the shortest direction then the span required for the grating will be 1000 mm less 300 mm, i.e. 700 mm. A suitable grating is 25 mm (1″) Duradeck I6000, which will support 7200 N/m² at this span with a deflection of less than 2.5 mm.

13.2.2 The longitudinal beams

These have a span of 3 m and are required to carry the live load of 4500 N/m², i.e. 6750 N per beam or 2250 N/m linear run.

Allow 225 N/m dead load. Then total beam load = 2475 N/m. Now a 150×150×6.25 mm wide flange beam in series 625 resin system [13.2] will carry 2500 N/m with a deflection of span/180, if the beam has lateral support to ensure transverse stability.

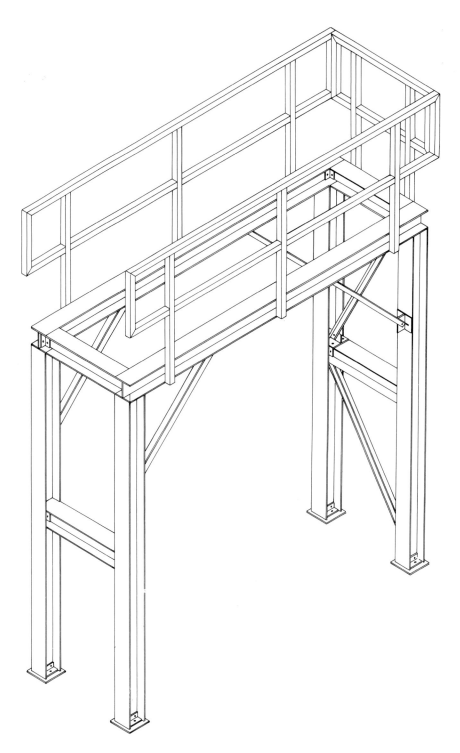

13.1 A composite platform over a chemical process tank.

13.2.3 The columns

Each column carries $(4500 \times 3/4)$ N = 3375 N plus, say, a dead load of 625 N (column and bracing) = 4000 N per column.

Assuming that both ends are pinned, then wide flange $100 \times 100 \times 6.30$ or 'I' $150 \times 75 \times 6.30$ are both adequate. However, connections are simplified by the use of wide flange $150 \times 150 \times 6.30$.

13.2.4 Sway bracing

As the platform is indoors there are no wind loads and sway bracing can be designed to provide stiffening in the transverse direction to resist 2% of the column load. Therefore 2% of 4000 N$\times$2 columns = 160 N.

$$\text{Axial load in the brace} = 160 \times \cos 63.4 = 72 \text{ N}$$

The cross-sectional area is 1200 mm. Therefore:

$$\begin{aligned} \text{stress} &= (72/1200) \text{ N/mm}^2 \\ &= 0.06 \text{ N/mm}^2 \end{aligned}$$

Consider back-to-back angles $50 \times 50 \times 6.30$. The allowable stress for the length of bracing (i.e. 2.65 m) is 8 N/mm^2 which is well in excess of the requirement and therefore suitable.

The knee braces will be adequate using similar sections.

13.2.5 Costs

The installed cost of this structure in GRP is highly dependent on the number of systems required. The example considered here is quite small and if only one is required then it is relatively expensive for a fabricator to set up and to undertake the job. However, if, say, 50 are required, or if the platform has an area of 150 m^2 rather than only 3 m^2 then significant cost benefits accrue. The savings are made in the three areas of cost: materials, fabrication and erection.

To produce one 3 m^2 GRP platform the costs are estimated as:

Materials	£1600
Fabrication	520
Erection	320
Total	£2440

To produce fifty 3 m^2 GRP platforms the costs of each are estimated as:

Materials	£1190
Fabrication	160
Erection	240
Total	£1590

13.3 Bibliography

[13.1] S. W. Tsai and H. T. Hahn, *Introduction to Composite Materials*, Westport, Technonic, 1980.

[13.2] *MMFG Design Manual*, Virginia, USA, Morrison Molded Fiber Glass Company, 1989.

[13.3] J. A. Quinn, *Design Manual, Engineered Composite Profiles*, Cheshire, Fibreforce Composites Ltd, 1988.

[13.4] ASTM D 3917-80, 'Tolerances for pultrusions'.

[13.5] ASTM D 3918-80, 'Definitions for pultrusions'.

[13.6] ASTM D 3916-80, 'Test methods'.

[13.7] ASTM D 3647-78, 'Classification of pultrusion'.

[13.8] G. Lubin, *Handbook of Fibreglass and Advanced Plastics Composites*, New York, Van Nostrand Reinhold, 1982.

[13.9] M. Holmes and D. J. Just, *GRP in Structural Engineering*, London, Applied Science, 1983.

[13.10] *AFC Corrosion Resistance Guide*, Runcorn, Cheshire, Fibreforce Composites, 1987.

14 GFRP minesweeper: Pre-production test structure with box core sandwich construction

14.1 History

No minesweepers were built between World War II and the 1960s and, because of the overwhelming use of glass reinforced plastics for boat building, skilled workers and facilities for making large wood boat structures had virtually disappeared. How could the Royal Navy in the early 1960s obtain a non-magnetic mine hunter? Would a glass fibre polyester structure be adequate to produce a hull structure many times larger than anything built so far and with more exacting design and loading cases? After years of extensive testing to determine the best glass fibre finish and resin system for long life and minimum deterioration in wet conditions, the MoD ordered Bristol Aeroplane Plastics Ltd to design, develop and build a full size test structure.

14.2 Description

The full size centre section representing a typical minesweeper was approximately 30 ft (10 m) beam, 30 ft (10 m) height and 50 ft (17 m) long. It was structurally complete except for bows, stern and superstructure. There was sufficient structure to demonstrate strength against sea service loading.

14.3 Design requirements

1 Corrosion resistance – long life.
2 Green sea pressures.
3 Wave loading creating overall shear and bending loads.
4 Equipment and engine support.
5 Deck loading.
6 Bulkhead flooding pressures.
7 Impermeability.

8 Shock loads – mine explosion.
9 Fire resistance.

14.4 Design approach

A sandwich structure was chosen to obtain minimum weight and make use of a newly developed fabric impregnation and dispensing system to obtain minimum cost. GFRP skins with metal honeycomb, foam or end grain balsa cores had already been demonstrated to produce efficient structures but unfortunately with critical weaknesses. For the three materials these can be listed as:

1 Metal honeycomb:
   (a) not suitable for a wet lay-up manufacturing system;
   (b) difficult to form double curvature;
   (c) honeycomb-to-honeycomb joint problems;
   (d) corrosion problems (Nomex honeycomb not then available);
   (e) expensive.
2 Foam: poor strength and dimensionally unstable, high creep characteristics, particularly at elevated temperatures.
3 End grain balsa: heavy.

A new patented sandwich core structure was available which had the advantage of creating an optimum balanced structure for GFRP skins over 6 mm thick. This comprised a box formed by hot pressing SMC thus forming a unique all-GFRP sandwich structure with good area of bond between the core webs and the skins.

The manufacturing sequence is shown in Fig. 14.1.

The skins were formed from woven roving fabrics for maximum strength and minimum weight and cost. Care was taken to ensure that the continuous glass fibres were in line with the direction of maximum loads or stresses. Unidirectional fabric was used to enhance the keel strength and stiffness.

Design of the bulkhead to hull, bulkhead to deck and deck to hull joints had special consideration requiring qualification by local structural tests.

The engine required local beams to spread the support loads into the hull. This was achieved by large foam fill top-hat sections with the foam as a non-structural manufacturing tooling aid.

One interesting aspect was possible water permeability of the hull especially if damaged. Should the box cores be sealed, or should they be open to allow bilge collection of water and thus ensure negligible pressure differential across the skins? Some boxes were open and some sealed to evaluate the difference.

Stage 1
Lay down the 'outer' skin.
Woven roving and polyester resin.

Stage 2
Inject epoxy adhesive into the
box core grooves.

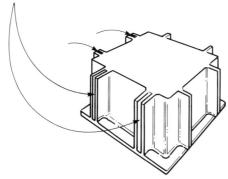

Stage 3
Imbed box core into skin (if wet)
or use epoxy adhesive.

Stage 4
Lay up 'inner' skin.

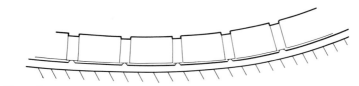

The photographs taken during construction (Fig. 14.2–14.14)
illustrate this process.

14.1 Manufacturing sequence for box core sandwich structure.

14.5 Structural analysis

All design loading requirements, with the exception of shock loading which will be discussed in the next section, were evaluated using typical aircraft structure procedures including shear flow analysis to design the forward and aft deck to superstructure intersections.

14.6 Shock loads

At the start of the project the shock resistance of the structure to mine explosions was only required to be as good as the current design of wood-constructed minesweepers. This was demonstrated by underwater explosive tests on panels, and it showed that the proposed box core sandwich structure was significantly stronger than wood.

As the test hull was nearing completion the USA released data on its proposed GFRP minesweepers. These included resistance to an underwater 'shock factor' of 3; this factor was defined as:

shock factor = weight of explosive/(distance)^3 lb/ft^3

consequently, this shock factor was adopted as a requirement for the ships of the United Kingdom. This change created project difficulties as a major design alteration had to be incorporated into the construction at a very late stage.

A sandwich structure is not the best structure to resist shock loads especially if the shock tests use small amounts of explosive at short distances. High pressure shock waves passing through the box core webs accelerate the inside skin to produce unacceptable bond line tensile loads. This was demonstrated by tests on large panels and the full scale hull section. Unfortunately the automatic lay-up system was so effective that the structure was translucent – this is illustrated in the photographs (Fig. 14.2–14.4) where the box cores in the bulkheads and cores in the hull can be clearly seen. Such a structure allowed interlaminar cracks caused by shock tests to be easily seen. Although the cracks were small, the effect was exaggerated owing to the clearness of the structure. Such defects would not have been observed in a normal hand laid-up structure and may not have been critical, this was demonstrated over a period of many years when attempts took place to destroy the structure. Owing to the frightening effect of the 'see through' cracks, a decision was made to abandon a box core design and adopt a thick single skin design which was reinforced with closely pitched stiffeners fastened by hundreds of titanium bolts to prevent shock wave separation. This produced a much heavier and more costly structure but nevertheless very successful; it has, after 20 years of excellent service, had its life expectancy increased, far in excess of the life of the equipment it carries.

14.2 Hull tool ready for lay-up.

14.3 Automatic resin impregnated glass fibre fabric dispensing system laying up the hull skins.

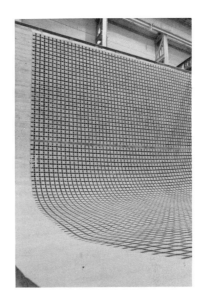

14.4 Assembly of box cores on hull structure.

14.5 Bulkhead joint structure.

14.6 Assembly of box cores to deck structure.

14.7 Deck sandwich structure waiting for second skin.

14.8 Skin lay-up of deck sandwich structure (note the translucent resin impregnation).

14.9 Reinforcing the hull.

14.10 Bulkhead assembly.

14.11 Deck assembly.

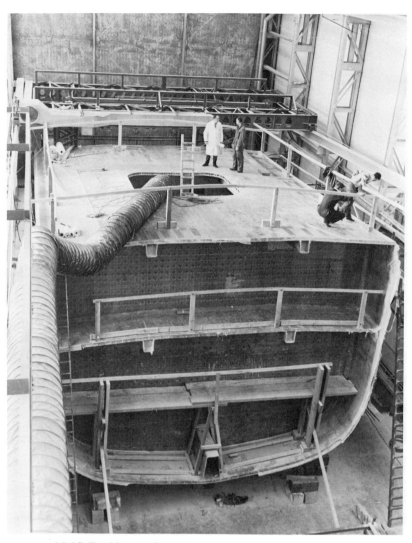

14.12 Final inspection.

14.13 Launch.

14.7 Manufacturing challenges

In the early 1960s Bristol Aeroplane Plastics had a deep conviction that progress in composite structures could only be made by the use of automated manufacturing processes, first to reduce costs and second to ensure a high quality product that was repeatable. This philosophy had been proved by successful work on resin injection moulding and, in particular, with filament wound structures. To this end, resin mixers, resin dispensers, resin–fabric impregnators, fabric laying machines and laminate consolidation equipment were developed for the construction of large flat and curved panels. Photographs taken during the building operation (Fig. 14.2–14.14) illustrate this system; they also show a gantry system for laying up the hull. With this equipment, which was unique, considerable maintenance problems were overcome, and a uniform and accurate resin content-controlled laminate was produced which was so well consolidated that it produced a translucent, almost void-free, structure. A significant aspect of this project was the fact that the test 'hull' had been built for only £4.50 per kg of the structure weight. This was an extraordinary feat for a 'first off' ultra-large component. This achievement was due to the enthusiasm and skill of the development team and also to the method of manufacture which avoided the need to use a tedious bucket and brush method of laying up the hull.

14.14 Ready for testing.

14.8 Test results

Static structural loading tests including hull bending and shear, and deck loading, were successfully demonstrated. Shock loading, a revised requirement after the structure was built, was repeated many times using underwater explosives. Although unacceptable interlaminar cracking of the sandwich hull structure occurred, no major structural damage was reported. Unconfirmed reports have stated that many attempts were made to destroy and sink the hull without success, including setting it on fire!

15 Shield for 4.5 inch Mark 8 gun manufactured by Vickers and used on Type 42 destroyers

This shield is a hemispherical structure, 4 metres in diameter. Production started in 1967, more than 45 have been made and they are still being manufactured.

15.1 Reason for composite materials

1 Low mass inertia for quick orientation of the gun platform.
2 Relative low cost because of difficulty of manufacturing a complex curved structure in metal.

15.2 Design requirements

1 Low inertia for maximum traverse setting speed.
2 Corrosion resistance.
3 Good shape – minimum profile.
4 Access doors.
5 Stiff structure to avoid movement between the shield mantlet opening and the gun barrel.
6 A sealed joint to split the shell for gun removal.
7 Temperature environment +51 °C, −25 °C.

Loading cases:

8 Nuclear blast: over pressures and reflected pressures.
9 Green sea wave impingement pressure loading.
10 Gun barrel blast pressure at mantlet opening – a fatigue case.
11 Shock loading.
12 Resistance to local damage and low energy fragments.

15.3 Design approach

A sandwich structure of polyester impregnated glass fibre woven roving skins using epoxy bonded end grain balsa wood core was chosen for:

1 Low weight stiffness to resist buckling from nuclear blast and wave pressure.
2 Ability to form the complex shield contours defined by the customer.
3 Use of different core densities and skin thicknesses for areas requiring different strength requirements.
4 Good impact, fatigue, temperature and dimensional stability resistance.

Resistance to local impact loads dictated minimum skin thicknesses of 6 mm on the outside and 4.5 mm on the inside.

15.4 Structural analysis

Two grades of end grain balsa wood were used. A stronger, heavier grade was used in the base joint, mantlet opening and other joint areas. In the general area of the shield, the lightest balsa grade available was used to resist buckling from external pressures.

The overall design aim was to achieve a fail-safe failure mode in the event of excessive impact loads or war damage by allowing sandwich core shear failure to occur first such that the skins would give continued protection to the gun.

Geometry and local material content was analysed and determined using ref. [15.1].

15.5 Tests

To demonstrate adequate material properties a number of tests were made determining flexural, tension, compression and interlaminar shear strength of the structure at room temperature and 70 °C, and after 72 hour water boil tests. Shear tests were made on the balsa wood blocks and syntactic foam to ensure that the desired design allowable strength could be met.

Tests on structures simulating possible manufacturing flaws, such as delays in fabric lay-up and joggles in the skin structure because of steps in the balsa core surface, were made to demonstrate that such a construction would still meet the minimum design properties.

Structural tests were also made on simulated critical parts of the shield to check local strength. Fatigue strength of mantlet opening resulting from muzzle blast was checked first on contoured test specimens and finally confirmed by gun firing tests on a production shield.

15.6 Manufacturing and challenges

1 Construction of tooling – especially the master model and mould tools with joints was one of the major manufacturing challenges requiring large scale precision to ensure assembly of parts.

2   Tooling to provide correct geometry for alignment of gun elevation was also critical to ensure freedom of movement.
3   Surface finish for long life.

Initially the surface consisted of glass tissue impregnated with the same polyester resin as that used for the woven roving skins but with added pigment to match the paint scheme. This method created blistering from voids under the tissue owing to inadequate lay-up skills. The majority of these blisters were exposed during transport of the shield from Bristol to Barrow-in-Furness. Such blisters and voids located by acoustic sensors were invariably repaired during assembly of the shield to the mounting platform prior to painting.

After 40 shields had been produced over a period of 18 years a spray gel coat surface was used with good effect, being the only shield not to have any void-filling actions after despatch. Unfortunately, spray gel coat was abandoned because of difficulties with maintenance of the spray equipment. The application of hand-applied gel coat continued with limited success owing to lack of application skills made difficult by difficult contours and size of the mould tool.

15.7 Quality assurance for structural reliability

To ensure that the structure was made with adequate strength, not only were pieces cut from each completed shield and tested, but in addition a large test panel was constructed in sequence with each manufacturing operation on the shield. When the gel coat was applied to the shield it was also applied to the test panel tool. In addition, as the outer skin was laid down this was also laid on to the test panel gel coat, followed by the core and so on until shield completion. After gelation of each stage a strip was cut off the test panel tool, cured, examined and tested. If any flaws or a low quality structure was found, it was then possible to stop shield production and remove the low quality material of the last operation. This method was effective in forcing operators to maintain maximum quality workmanship and to ensure that shields were completed with no major quality problems.

15.8 Service experience

After the Falkland Island war a request was made to strengthen the main door. To aid design, a door was returned that had been damaged during the conflict. An examination indicated that the damage was that which would occur if the door had been left open with the door handle levers closed. A South Atlantic gale would have banged the door open and shut many times, impacting the structure on the closed door handles.

Despite this rough treatment the door was still functional. Parts of the damaged and split structure were examined and found reasonably

15.1 Type 23 Frigate gun shield.

sound. The skins were intact with no signs of deterioration except for bruising from impact of the door levers. The balsa core had sheared and split and in some areas, where exposed, had absorbed water: but there was no sign of rotting. A damaged section is shown in Fig. 15.7.

A redesigned, stronger, door was supplied for subsequent shields and to replace the damaged one, but no request was made to replace any existing door. In fact no shield damage other than the single door has been reported. A good case history considering the long severe service life.

15.9 References

[15.1] S. Timoshenko and S. Woinowsky-Krieger, *Theory of Plates and Shells*, New York, Toronto, London, McGraw-Hill, 1959.

15.2 Final assembly.

15.3 Assembly line.

15.4 Assembling the joint.

15.5 Rear view.

15.6 Loading for delivery.

15.7 Section of 'storm'-damaged door.

Index

Aberfeldy bridge, 243
accelerator, 8, 9, 60
acetyl acetone peroxide, 9
acid amides, 25
additives to resins, 21
adherends
 non-identical, 217
 tapering, 217
adhesive
 brittle characteristics, 211
 characteristics in joints, 210
 ductile characteristics, 211
 shear stress distribution, 218
 strengths, 218
aliphatic
 amines, 67
 hydroxyl, 17
alkylamine, 25
alpha-methyl styrene, 7
aluminium honeycomb, 131
amine
 curing agents, 18
 resin, 18
angle-ply laminates, 130, 138
anhydrides, 18
anisotropic
 polymer composites, 118
 properties of composites, 108
Anmac GFRP Fabricators, 276
ANSYS computer program, 178, 296
antistatic agents, 25
aromatic
 amines, 67
 groups, 17
artificial intelligence system (ATC), 301
autoclave, 85
auto-spray, 80

beam structures – thin walled, 148
bisphenol
 'A' 16
 resin, 7, 12
benzol peroxide, 9
brominated resins, 17
bolted joint, 226
 design, 239

bolted-bonded joint, 240
boron fibres, 33
brittle fracture modes of composite, 192
British Standards Specification (476),
 263–271, 273, 278
Buckling
 failures in box beams, 160
 instability, 192
 parameter, 160
buffing composites, 101
building regulations (Class 0) fire rating,
 263, 266, 274, 275, 279

computer aided design, 296
carbon black, 23
carbon fibre, 5, 23, 31, 131, 73
 manufacture of, 31
 mechanical properties of, 31
 reinforcement, 68
carbonates
 calcium, 21
 magnesium, 22
cast resin, properties of, 19
catalyst, 8, 66
C E mark, 274, 275
centrifugal casting technique, 81
characterisation of FRP materials 186,
 189
cleaning solvents, 66
closed mould process, 83
 cold press, 86–88
 hot press, 87
 injection moulding, 92
 pressure bag, 85
 pultrusion, 94
 resin injection, 89
 vacuum bag, 83
co-polymerisation reaction (curing), 7
COMPOSIC FEA program, 178
composite
 beam design formulae, 151
 materials
  anisotropic, 108
  characteristics, 122
  design considerations, 107
  isotropic, 108

panels – design of, 140
properties, variation with temperature
of, 43
stacking sequence, 213
cone calorimeter, 275, 279
continuous
filament winding, 98
laminating process, 94
cost – design for, 110
creep design coefficients, 121
cross-ply laminates, 128, 138
cross-link density resins, 12
cure of polyester resin, 9
curing agents, 66
cutting and trimming composites, 99
cycloaliphatic amines, 67
cyclohexanone peroxide, 9
CYLAN PC computer program, 163, 177

Design
analysis – preliminary, 136
data sheets – geometric section
properties, 152
management, 111
material data for polymer composites,
119
process, 111
dibasic acid (unsaturated), 7
Diglycidyl, 17
DMC, 32
doctor blade, 94
dough moulded compound, 86
drilling composites, 100

environment – designing for, 109
environmental factors 186
epichlorohydrin, 4, 16
epoxy resins, 4, 16–18, 25, 32, 131
cold cure, 4
properties of, 19
exothermic reaction, 9
ESDU, 210, 251, 252
etching composites, 101
European market, 274
European Standardisation Committee
(CEN), 181

fasteners – countersunk, 238
fatigue, 186, 193
failure
loads
in-plane tensile, 139
in-plane shear, 139
modes and effect analysis (FMEA),
301
modes – joints, 207
stress – average laminate failure stress,
139
Falkland war, composite service in, 327

fibres, see carbon fibres and glass fibres
aramid, 73
polyaramid, 5
filament
winding, 79, 80
wound vessels – design of, 173
fillers, 69
fillite, 23
finishing methods for composites, 99
fire
performance, 263
propagation, 263
index, 276
retardant fillers, 9, 21
retardation grade polymer, 315
tests, 266
development, 173
spread of flame, 265, 271
foam cores, 69
furane resins, 20, 68

galvanic corrosion, 201
gel coat
hand applied, 74
spray, 327
glass-fibre, 23, 24, 73
A-glass, 24
C-glass, 24
chopped strand mat, 3, 4, 27, 43, 44, 58
commercially available, 26
continuous
filament mat, 29
filament rovings, 27
E-glass, 4, 24, 27
filament, 25
manufacture of, 24
properties of, 26
R-glass, 4, 24
S-glass, 4, 24
woven
cloth, 3
glass fabrics, 29
rovings, 4, 29, 44, 58
reinforcements, 68
satin weave, 29, 30
twill weave, 29
unidirectional weave, 30
glass
flakes, 22
spheres – hollow, 22
tissue, 327
glycidyl ethers, 17
GRP minesweeper, 309

hand lay-up technique, 74
health and safety, 63
ingestion, 64
ingestation of phenolic resin solutions,
68

inhalation, 64
heat deflection temperature (HDT), 9, 12, 21
Heathrow terminal, 2, 262, 276
HET acid resin, 276
hot press moulding technique, 87

injection moulding, 92
interface – fibre/resin, 5
ISO-DIS 5661 (international standard test), 275
ISO-NPG, 7, 12
isophthalic resin, 7, 9, 12, 43, 248
  gel coats, 248
isotropic composite material, 108, 162, 164

Johnson, A.F., 129
joints
  adhesive selection, 205
  bolted joint design, 239
  bolting, 103
  bonded, 102, 203
  design, 109, 211
    of single lap joints, 212
  dimensions, 207
  environmental effects, 238
  efficiency, 212, 220
  failure modes, 207, 227
  fastener selection, 226
  fatigue loading, 238
  geometric ratios, 229
  geometry, 206, 227
  lap configuration, 239
  mechanical – bearing strengths of, 234
  mechanically fastened joints, 223
    hole preparation, 224
  mechanism of load transfer, 202
  multifastener joints, 235
  overlap length, 205
  rivets, 102, 234
  scarf joints, 222
  single bolt hole, 229
  stepped joint design, 220
  surface preparation for adhesive, 205
  types of, 202
jute, 23
  fibrous, 32

kieselguhr, 22
Kevlar fibres – mechanical properties, 119, 120

LAMICALC computer program, 177, 178
LAMINA computer program, 284
LAMPCAL computer program, 147
laminate material characteristics, 229
law of mixtures, 44
leaky mould technique, 86
life performance – design for, 110

limit state design, 181, 182, 189, 283
  serviceability, 190
  ultimate, 190
long term loading, 120, 186
  load reduction factors, 122

manufacture – design for, 110
manufacturing challenges to boat building, 322
Marco, 3
MCMV vessel, 245
MEKP, 9
Mercapto, 25
metal
  oxide fillers, 22
  powder fillers, 22
methacryloxy, 25
methodology and management of design projects, 107–115
methyl methacrylate, 7
MIC-MAC PC program, 162, 177, 178
mineral fillers, 21
monomer, 6
Motor Industries Research Association (MIRA), 289

napkin-ring test, 210
netting analyses, 131
Nomex, 246
  core, 260
nylon fabrics, 32

Ogorkiewicz R M, 121
open mould process, 74
  auto spray, 80
  centrifugal casting, 81
  filament winding, 80
  hand laminating, 74
  spray laminating, 76
  winding, 80
orthophthalic resins, 7, 12
orthotropic composites, 107, 108
  properties, 107, 108
  laminate
    inplane strains, 123
    bending and twisting curvatures, 123
    specialty, 123

painting composites, 101
PANDA computer program, 146
partial coefficients – methods of, 183–185, 189
partial material coefficients, 121, 184, 191, 286
P/COMPOSITE FEA program, 178
peel ply, joints, 205
PERMASLA FEA program, 178
permeability of sandwich construction, 310

peroxides, 8, 9
phenolic
  foam, 131
  (phenol-formaldehyde) resin, 3, 20, 68
  resin – ingestion of, 68
pigments, 23, 69
plaining composites, 100
polyamide curing agents, 67
polyaramid
  fibres, 5, 23, 30, 31, 33
  manufacture of, 30
  reinforcement, 68
polycarbonate, 20
polyester
  fibres, 32
    properties of, 32
  resin (unsaturated), 4, 6, 7, 12, 16, 23, 25, 43, 65
    properties of, 13
polyethersulphone, 4
polymer composite material characteristics, 122
polypropylene, 20
polyimide resin, 20
polystyrene, 6, 20
  foam, 131
polyurethane foam core, 69, 78, 131
  integral skin foam, 131
polyvinylchloride foam core, 69, 78, 131, 260
prediction of mechanical properties of polymer composites, 43
preliminary design analysis for polymer composites, 136
pressure
  bag technique, 85
  loaded plates – design parameters, 144
pultrusion
  profiles, 284, 285, 303
  technique 97
pumice, 22

quadratic stress criterion, 127
quality assurance, 327

radomes, 3
randomly orientated polymer composites, 118
reliability, 182
resin – HET acid, 276
  transfer moulding (RTM), 29, 293, 296, 297, 299, 300
riveted joints, 234
routing –high speed – for composites, 100
Rover 800 rear beam, 292, 293

SAMCEL FEA program, 178
sandwich beam

critical buckling load of, 132
  design of, 167
  low shear modulus core, 132
  material, design of, 167
  safety, degree of, 182
  stiff shear modulus core of, 134
  wrinkling instability of, 133
sandwich construction, 56, 131, 165
  compressive failure load of, 166
  core crimping of, 166
  correction factor for, 168
  failure load of, 166
  flexural rigidity of, 60
  for minesweeper, 310
  materials of, 131
  properties of materials for, 164
  shear stiffness of, 60
  shear strength of, 59
  stiffness properties of, 165
  strength properties of, 165
  tensile failure load of, 166
  thick face, 132
  thin face, 57–63, 132–134, 165–168
  transverse shear stiffness of, 165
sandwich panels
  classification, 132
  design of, 169
  thick face, 132
  thin face, 132
sandwich
  strut, 133
    design of, 165
sawing composites, 99
Scott Bader, 3, 276
self-polymerisation, 18
shear
  correction factor, 151
  material failure, 142
sheet moulding compound, 88
shield for Mark 8 gun, 325
sideguard system, 281, 283, 289
silica, 22
  aerogel (fumed silica), 23
silicate, 22
  calcium aluminium, 22
  magnesium, 22
  zirconium, 22
silicon-organo, 25
silicone resins, 20, 25
sisal fibres, 32
slabstock foam, 69
spray-up technique, 76
spray winding technique, 80
stress theory of failure – maximum, 125
structural characteristics of FRP material, 186
styrene, 6, 15
  monomer, 12
surface finishes for composites, 101

Terminal 2, Heathrow Airport, 262, 276
thermoplastic resins, 20
thermosetting resins, 20, 23, 73
thick adhesive test, 210
thin-walled cylindrical vessels,
  closed, 172
  open, 172
thin-walled vessels, design of, 171, 173
tips for beam design, 159, 162
Tsai-Hill criterion, 127
transverse ply failure, 129

urethane-acrylate resin, 12, 13
ureol tool, 297, 298
U-V radiation, 7, 21, 276
ultimate limit state, 183, 190

vacuum
  assisted resin injection, 91
  bag technique, 83
vermiculite, 22
Vetrotex unifilo, 296
Vickers gun shield, 325
vinyl alkylamine, 25
vinylester resins, 13, 16
  casting of, 13
  curing of, 13
vinyl toluene, 7

Williams, J.G., 121

Young, W.C., 152

zirconium silicate, 22